TERMITES
Biology and Pest Management

Quotations

He of mighty energy assumed the posture called Vira
quiet and still like an inanimate post,
and for a long period remained at the same spot of ground.
And he was turned into a termite mound
covered over with creepers.
And after a long period,
swarms of termites enveloped him.
And covered all over with termites,
the sagacious saint looked exactly like a heap of earth.
Book of the Vana Parva
Chapter 123
McKevan (1978)

When Solomon died knowledge of his death
would have caused the jinn to stop their work,
the corpse was made to stand leaning on a staff,
causing the illusion that Solomon was alive,
and a long time later 'termites' ate the staff away
causing the corpse to fall down that the jinns
realised that Solomon had died by then.
Qur'an (XXXIV. 13)

Termites
Biology and Pest Management

M.J. PEARCE

formerly of the Natural Resources Institute
Chatham, Kent, UK

CAB INTERNATIONAL

CABI *Publishing* is a division of CAB *International*

CABI Publishing
CAB International
Wallingford
Oxon OX10 8DE
UK

Tel: +44 (0)1491 832111
Fax: +44 (0)1491 833508
Email: cabi@cabi.org
Web site: http://www.cabi.org

CABI Publishing
10 E 40th Street
Suite 3203
New York, NY 10016
USA

Tel: +1 212 481 7018
Fax: +1 212 686 7993
Email: cabi-nao@cabi.org

A catalogue record for this book is available from the British Library, London, UK.

Library of Congress Cataloging-in-Publication Data
Pearce, M.J.
Termites : biology and pest management / M.J. Pearce.
p. cm.
Includes bibliographical references (p.) and index.
ISBN 0-85199-130-0 (alk. paper)
1. Termites. 2. Termites—Control. I. Title.
QL529.P43 1997
595.7'36—dc21 97-13544
CIP

ISBN 0 85199 130 0

First printed 1997
Reprinted 1999, 2000

Typeset in 10/12pt Optima by Columns Design Ltd, Reading.
Printed and bound in the UK at the University Press, Cambridge.

Contents

Foreword	**vii**
Preface	**ix**
Acknowledgements	**xi**
1. Termites as Insects	**1**
What is a Termite?	1
Evolution	2
Relationship to Cockroaches	7
Castes of Termites	9
Classification of Termites	19
2. Distribution	**32**
World Distribution	32
Pest Distribution	33
Factors Affecting Distribution	37
3. Termite Biology and Behaviour	**40**
Communication	40
Feeding	46
Water Requirements	53
Defence	55
Foraging	60
Nest Building	63
4. Nest Systems	**65**
Nest Types	65
Termitophiles	74

5. Termite Ecology **77**
Soil Type 77
Vegetation Types 78
Benefits to the Environment 79
Environmental Factors 81
Predators and Parasites 85
Other Uses of Termites 90

6. Termites as Pests **91**
Food Preferences 91
Damage Recognition and Detection 93
Damage Assessment 100

7. Control Methods **101**
Chemical Control 101
Physical and Cultural Control 114
Biological Control 117
Safety 120
Future Control 121

Appendices **123**
1. Collection and Identification 123
2. Culture Methods 127
3. Monitoring Methods 137
4. Laboratory Tests Using Termites 143

References **153**
General References 153
Taxonomic References 154
References on Biology and Control 157

Index **169**

Colour plates

Foreword

Mike Pearce has undertaken a formidable task. There are well over 2000 different species of termites in the world. They are cryptic insects, nesting in mounds as hard as concrete, in mile-long tunnels deep in the soil, or in galleries carved within solid wood. Although they have specialized to feed upon a resource (cellulose) where they have few competitors, they do so in a variety of different ways. Some termites feed exclusively on dead wood, such as the lumber in our homes, with the assistance of symbiotic protozoa and bacteria that can only live for minutes outside their termite partners. Others feed on living trees or crops, some moving openly across the tree limbs or the soil surface and others constructing mazes of covered paths from masticated plant material and collected soil particles. The most advanced termites are gardeners, cultivating and carefully tending species of fungi in their mounds that grow nowhere else on the face of the earth.

Most significantly, termites are social animals. Unlike the ants and bees, where females do the work and males live fast and die young, male and female termites work side by side to tend their queen, feed the young, and protect the colony from ants and other invaders. The mechanisms which suppress sexual development of these individuals in the presence of the king and queen, and which stimulate their development into soldiers and other specialized castes, are still poorly understood, even though the termite 'queen pheromone' was one of the original hypothetical examples put forth when the term pheromone was coined in the 1950s. Chemistry is also known to play an important part in both termite communication and defence. Termite chemical defences are particularly varied, ranging from glues to strong irritants, and even to noxious chemicals released from exploding abdomens.

The termite way of life is truly unique in the insect world, and the expressions of that way of life by the over 2000 species of termites are incredibly diverse. In the first five chapters of this volume, the author has done an excellent job of explaining that way of life and capturing their diversity. Those who have thought of termites only as pests, if they thought of them at all, will be enthralled by the complex social and behavioural interactions and ecological relationships laid out before them. Specialists will find the comprehensive generic classification and the pictorial keys to be solid contributions, and extremely useful to those of use who frequently encounter immigrant termites.

Of course, termites are not only fascinating animals but are indeed serious pests of trees, structures and wood products. The final two chapters deal with this dark side of the termite story. The author's experience with termites in the tropics adds an interesting emphasis to this discussion for readers who have dealt with termites only in the temperate and more developed nations of the world. In the past decade, termite control world-wide has changed significantly with the decline in use of persistent organochlorine soil insecticides. Thus, in addition to describing the new generation of chemical control agents, Mike Pearce has put appropriate emphasis on physical control methods, termite baits and exploratory research into microbial control.

Finally, the appendices provide an excellent summary of various ecological and laboratory techniques commonly used by termite specialists. Students and readers involved in pest control rather than academia should appreciate this collection of both simple and more complicated methods. Throughout the text, the reader is provided with key references that will lead the more curious into the technical literature in each area.

Mike Pearce has not only undertaken a formidable task, but has completed it in an exemplary fashion with an unique international flavour. Welcome to the world of termites!

J. Kenneth Grace
Professor of Entomology
College of Tropical Agriculture and Human Resources
University of Hawaii at Manoa
USA

Preface

Termites, like ants, bees and wasps, are social insects, each individual acting as part of a group governed by the demands of the colony and ultimately by the queen. The word 'termites' comes from the Latin word *tarmes* which was given to a small worm that makes holes in wood. The termite society in some species has reached the same level of sophistication as the more advanced Hymenoptera. Termite queens can produce from a few hundred to more than 10 million eggs per year in some species, resulting in very large colonies. These are highly organized, relying on chemical and sensory messages for communication and defence, enabling them to exist in total darkness. Their differences from Hymenoptera, such as ants, bees and wasps, have resulted in them being placed in a separate order, the Isoptera.

Termites have an important place in economic entomology, with the cost of damage to buildings, especially in developed countries in America and Asia, amounting to many millions of pounds. Damage to houses by termites can, in some countries, exceed that caused by natural disasters and fires in a single year. In developing countries they have even more impact, destroying local huts and crops of poor subsistence farmers. Villages in India and Egypt have been destroyed by termites and the inhabitants forced to move to other areas. In Asia, ancient temples have also been attacked.

Much of the past history of humankind has been lost due to termites eating books, buildings and even destroying cave paintings and archaeological remains. Any material that incorporates cellulose can be devoured by termites, from paper to palaces and fungi to fir trees. The first indication of termite infestation often comes too late, when doors and roofs fall down or floors collapse. Termites, in their quest for food, also destroy other material that stands in the way; this has led to breaks in dam linings, fires and electrical faults in large cables.

Termite behaviour is instinctive and often opportunistic. They are able to create their own environment in concealed chambers in wood, or by building ventilated nests on or under soil or trees. Nests of ground-dwelling termites can extend to several metres underground, and have even been found in a twenty-two-storey apartment and seaworthy boats. Apart from causing destruction, some species of these blind builders can construct amazing cathedral-like nests using tons of soil. In proportion, human constructions appear trivial.

Termites are an important part of the food chain for many animals including humans. A nest can be owned by a farmer and the termites produced represent a valuable source of protein for the whole family. Mounds can provide shelter and homes for many animals and therefore help with survival and community ecology. Some termite mounds can provide a richer soil type as a result of termite activity, thus increasing crop production and changing vegetation patterns. Soil from mounds can be used by humans for pots, houses and even roads. In some African tribes termites can act as oracles. A piece of wood placed in a mound can help to solve important decisions depending on whether it is eaten or not eaten. The attack on different kinds of wood can also be used to make decisions where several different choices have to be made.

Several excellent detailed books have been produced on aspects of the biology and control of termites (see General References). This book aims to give an overall picture of the major aspects of termite biology, ecology and control that will give the student and the research worker a basic insight into the termite world, with emphasis on some of the new developments in biology and safer methods of pest management. Control of termites has now become a major problem as the more persistent organochlorine insecticides are now banned in many countries and control may require the use of new and unfamiliar equipment and techniques for many people. Damage to buildings, crops and forestry is increasing, since effective replacements are too expensive.

The appendices illustrate several practical methods, including collection, identification, culture and some types of laboratory tests. A selective bibliography is also included. The first section is General References on termites with bibliographies whilst the second is Taxonomic References. References on Biology and Control is the third section, but the emphasis here is on ecology and pest management as many of the more general references on biology can be found by consulting the works in the General References.

Acknowledgements

In preparing this book I owe a debt of gratitude to all those who have worked on termite research over the last twenty years. One can fill several volumes on references on termites as has been done in many bibliographies. I have tried to include many of the current works on termites which should give the reader a source of previous work by other researchers.

I am grateful to all those who have sent me information and photographs for this book, and especially to the Natural Resources Institute (NRI), Chatham, and the staff in the biology department for access to information sources on termites, and help over the years. I thank my past colleagues Dr Tom Wood, Jim Logan, Solomon Bacchus and Brian Waite for their help and use of photographs. I also thank especially Dr Jo Darlington (University of Cambridge, UK) for her comments on the draft relating in particular to the biology/ecology of termites and Professor Ken Grace (University of Hawaii at Manoa) for his valuable comments and update on biology and current methods of pest management in Hawaii and the USA, and also for writing the foreword to this book. Thanks are also due to the Department for International Development (formerly the Overseas Development Administration), UK, for providing financial assistance towards the cost of printing the colour plates. My special thanks are extended to my family for enduring my absences while writing this book and especially my wife for proofreading and encouragement, and my parents for their support.

SOURCES OF PLATES

Natural Resources Institute (NRI), Chatham, UK	1–3, 6, 10–14, 22, 26, 32
Dr J.E. Ruelle, Museum of Central Africa, Belgium	7
Dr T.G. Wood (NRI)	8
Bob Wright, Wrightway Pest Control, Florida, USA	18, 28
Prof. N.-Y. Su, University of Florida, USA	19
Dr B.M. Ahmed, CSIRO, Australia	29, 31
Dr Xia Chuan-Guo, Guangdong Entomological Institute, China	9
Dr Z. Salihah, Nuclear Institute of Food Agriculture, Pakistan	5, 16, 25

SOURCES OF FIGURES

Dr J.E. Ruelle, Museum of Central Africa, Belgium	5, 37, 39
Natural Resources Institute, Chatham, UK	1, 23, 28, 35, 57
Prof. R. Leuthold, University of Bern, Switzerland	26, 58
Dr S. Bacchus (ex NRI)	29
Dr J.A.H. Benzie, Australian Institute of Marine Science	35
Dr J. P.E.C. Darlington, University of Cambridge, UK	36
Dr R.H.L. Disney, University of Cambridge, UK	40, 41
Prof. G.D. Prestwich, University of Stony Brook, USA	43
Dr C.H. Sammaiah, Kakatiya University, India	45
Dow Elanco, Indianapolis, USA	51
Dr K. Tsunoda, Faculty of Agriculture, Kyoto Universty, Japan	52, 53

Chapter 1

Termites as Insects

WHAT IS A TERMITE?

Termites, just as some ants, have castes such as workers, soldiers and winged reproductives (Fig. 1). Due to their poor abdominal sclerotization, especially in the worker caste, they can appear white and are often called 'white ants' in some parts of the world. There are, however, many differences between ants and termites.

One of the major differences between termites and other social insects, such as ants, bees and wasps, is that the latter have larval and pupal stages that are not active within the colony. The word 'larvae' is commonly used in this book to mean very young termites, but it must be noted that even though termites are immature they are not really 'larvae' in the true sense as they do not enter a pupal stage as holometabolous insects do but appear as miniature versions of the adults as is the case with cockroaches and locusts. Another difference between other social insects and termites is that the male termite (king) remains with the female throughout her lifetime and does not die after mating. Ants also have a thin waist and elbowed antennae, which termites do not possess.

Termite reproductives have wings of equal size, which is where the Order name 'Isoptera' originates (*Isos* in Greek meaning the same and *ptero* meaning wings). In ants, the forewings are usually larger. There are seven families of termites in the order Isoptera, three of which are divided into sub-families. Members of the family Termitidae are often termed the 'higher' termites as they possess more advanced features. The other families are termed 'lower' termites. A list of genera for all families with authorities (authors who have published a description) is given in Table 1.

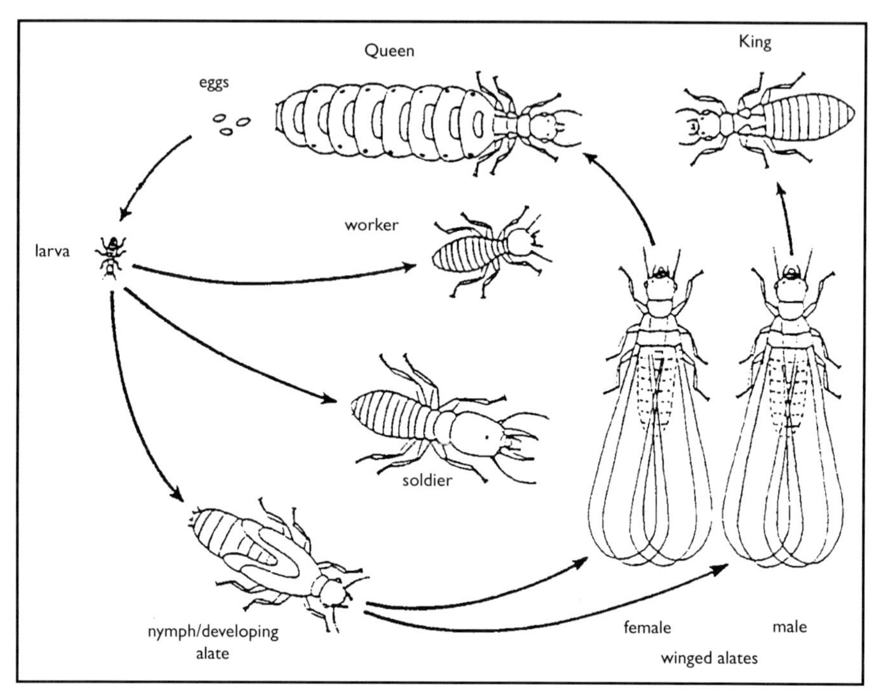

Fig. 1. Examples of the different castes of termites (Kalotermitidae).

EVOLUTION

Termites have been around on this planet for over 100 million years. They were in existence, therefore, before flowering plants. Cockroaches were present over 250 million years ago, while Isoptera (termites) could date back to Mesozoic or late Palaeozoic times (Table 2). Studies of fossils of petrified forests in Arizona, however, suggest that termites existed 220 million years ago, that is 100 million years before any other social insect. Previously, the oldest evidence of an actual fossil termite recorded was *Uralotermes permianum* from the Ural mountains, but as the fossil consisted of a single wing there is some uncertainty in dating this back to the Palaeocene period.

Another termite, *Valditermes brenanae* Jarzembowski, possibly from the family Cretatermitinae, of around 120 million years old was found in southeast England (Jarzembowski, 1981). Also in this family *Cretatermes carpenteri,* a member of the sub-family Hodotermidae, has been found. This was from the mid-Cretaceous period and identification was based on a forewing. A record of a Cretaceous termite, from Brazil, is *Spargotermes costalimai,* which belongs to the family Mastotermitidae. Also from Brazil,

Table 1. A generic classification of termites with authorities and geographical regions.

ISOPTERA Brullé, 1932					
MASTOTERMITIDAE Desneux, 1904					
Mastotermes (6)					
Froggat, 1896					
KALOTERMITIDAE Banks, 1919					
Allotermes (3)	*Ceratokalotermes* (6)	*Eucryptotermes* (5,8)	*Marginitermes* (7)	*Procryptotermes* (2–6, 8)	Tauritermes (8)
Wasmann, 1910	Krishna, 1961	Holmgren, 1911	Krishna, 1961	Holmgren, 1980	Krishna, 1961
Bicornitermes (2)	*Comatermes* (8)	*Glyptotermes* (1–6, 8)	*Neotermes* (2–6, 8)	*Proneotermes* (8)	
Krishna, 1961	Krishna, 1961	Froggatt, 1896	Holmgren, 1911	Holmgren, 1911	
Bifiditermes (2–4, 6)	*Cryptotermes* (1–8)	*Incisitermes* (4–8)	*Paraneotermes* (7)	*Pterotermes* (7)	
Krishna, 1961	Banks, 1906	Krishna, 1961	Light, 1937	Holmgren, 1911	
Calcaritermes (4, 7, 8)	*Epicalotermes* (2–4)	*Kalotermes* (1–4, 6–8)	*Postelectrotermes* (1–4)	*Rugitermes* (5, 8)	
Snyder, 1925	Silverstri, 1918	Hagen, 1853	Krishna, 1961	Holmgren, 1911	
TERMOPSIDAE Grassé, 1949					
TERMOPSINAE Holmgren, 1911	**POROTERMITINAE** Emerson, 1942	**STOLOTERMITINAE** Holmgren, 1909			
Archotermopsis (4)	*Porotermes* (2, 6, 8)	*Stolotermes* (2, 6)			
Desneux, 1904	Hagen, 1858	Hagen, 1858			
Hodotermopsis (4)					
Holmgren, 1911					
Zootermopsis (7)					
Emerson, 1933					
HODOTERMITIDAE Snyder, 1925					
Hodotermes (1, 2)	*Microhodotermes* (1, 2)	*Anacanthotermes* (1, 2, 4)			
Hagen, 1853	Sjöstedt, 1926	Jacobson, 1904			

Continued

Table 1 *Continued*

RHINOTERMITIDAE Light, 1921

COPTOTERMITINAE	**HETEROTERMITINAE**	**PSAMMOTERMITINAE**	**TERMITOGETONINAE**	**STYLOTERMITINAE**	**RHINOTERMITINAE**	**PRORHINOTERMITINAE**
Holmgren, 1910	Froggat, 1896	Holmgren, 1911	Holmgren, 1909	K&N Holmgren, 1917	Froggat, 1896	Quennedey and Deligne, 1975
Coptotermes (1–8)	*Heterotermes* (2–8)	*Glossotermes* (8)	*Termitogeton* (4)	*Stylotermes* (4)	*Acorhinotermes* (8)	*Prorhinotermes* (3–6, 8)
Wasmann, 1896	Froggatt, 1896	Emerson, 1950	Desneux, 1904	Holmgren	Emerson, 1949	Silvestri, 1909
	Reticulitermes (1, 4, 7)	*Psammotermes* (1–4)		and Holmgren 1917	*Dolichorhinotermes* (8)	
	Holmgren, 1913	Desneux, 1902			Snyder and Emerson, 1949	
	Tsaitermes (4)				*Macrorhinotermes* (6)	
	Li and Ping, 1983				Holmgren, 1913	
					Parrhinotermes (6)	
					Holmgren, 1910	
					Rhinotermes (8)	
					Hagen, 1858	
					Schedorhinotermes (2, 4–6)	
					Silverstri, 1909	

SERRITERMITIDAE Emerson, 1965

Serritermes (8)
Wasmann, 1897

TERMITIDAE Light, 1921

MACROTERMITINAE Kemner, 1934

Acanthotermes (2)	*Ancistrotermes* (2)	*Megaprotermes* (2)	*Odontotermes* (2, 4)	*Pseudacanthotermes* (2)	*Synacanthotermes* (2)
Sjöstedt, 1900	Silverstri, 1912	Ruelle, 1978	Holmgren, 1912	Sjöstedt, 1924	Holmgren, 1910
Allodontermes (2)	*Macrotermes* (2, 4)	*Microtermes* (2–4)	*Protermes* (2)	*Sphaerotermes* (2)	
Silvestri, 1914	Holmgren, 1909	Wasmann, 1902	Holmgren, 1910	Holmgren, 1912	

APICOTERMITINAE Grassé and Noirot, 1954

Acholotermes (2) Sands, 1972
Acidnotermes (2) Sands, 1972
Acutidentitermes (2) Emerson, 1959
Adaiphrotermes (2) Sands, 1972
Aderitotermes (2) Sands, 1972
Adynatotermes (2) Sands, 1972
Aganotermes (2) Sands, 1972
Allognathotermes (2) Silvestri, 1914
Alyscotermes (2) Sands, 1972
Amalotermes (2, 7) Sands, 1972
Amicotermes (2) Sands, 1972
Anaorotermes (2) Sands, 1972
Anenteotermes (2) Sands, 1972
Anoplotermes (7, 8) Müller, 1873
Apagotermes (2) Sands, 1972
Aparatermes (8) Fontes, 1986
Apicotermes (2) Holmgren, 1912
Asagarotermes (2) Sands, 1972
Astalotermes (2) Sands, 1972
Astratotermes (2) Sands, 1972
Ateuchotermes (2) Sands, 1972
Coxotermes (2) Grassé and Noirot, 1954
Doonitermes (4) Chatterjee and Thakur, 1966
Duplidentitermes (2) Emerson, 1959
Eburnitermes (2) Noirot, 1966
Euhamitermes (4) Holmgren, 1912
Eurytermes (2) Wasmann, 1902
Firmitermes (2) Sjöstedt, 1926
Grigiotermes (8) Matthews, 1977
Heimitermes (2) Grassé and Noirot, 1954
Hoplognathotermes (2) Silvestri, 1914
Indotermes (4) Roonwal and Sen-Sarma, 1958
Invasitermes (6) Miller, 1984
Jugositermes (2) Emerson, 1928
Labidotermes (2) Deligne and Pasteels, 1969
Machadotermes (2) Weidner, 1974
Orientotermes (4) Ahmad, 1976
Phoxotermes (2) Collins, 1977
Rostrotermes (2) Grassé, 1943
Ruptitermes (8) Matthew, 1977
Sinotermes (4) He and Xia, 1981
Skatitermes (2) Coaton, 1971
Speculitermes (4) Wasmann, 1902
Tetimatermes (8) Fontes, 1986
Trichotermes (2) Sjöstedt, 1924

TERMITINAE Sjöstedt, 1926

Ahamitermes (6) Mjöberg, 1920
Amitermes (1, 2, 4–8) Silvestri, 1901
Angulitermes (1, 2, 4) Sjöstedt, 1924
Apilitermes (2) Holmgren, 1912
Apsenterotermes (6) Miller, 1991
Basidentitermes (2) Holmgren, 1912
Batillitermes (2) Uys, 1994
Capritermes (3) Wasmann, 1897
Cavitermes (8) Emerson, 1925
Cephalotermes (2) Silvestri, 1912
Crenetermes (2) Silvestri, 1912
Cristatitermes (6) Miller, 1991
Cubitermes (2) Wasmann, 1906
Cylindrotermes (8) Holmgren, 1906
Dentispicotermes (8) Emerson, 1949
Dicuspiditermes (4) Krishna, 1965
Dihoplotermes (8) Araujo, 1961
Drepanotermes (6) Miller, 1991
Ephelotermes (6) Miller, 1991
Ekphysotermes (6) Gay, 1971
Foraminitermes (2) Holmgren, 1912
Forficulitermes (2) Emerson, 1960
Furculitermes (2) Emerson, 1960
Genuotermes (8) Emerson, 1950
Globitermes (4) Holmgre, 1912
Gnathamitermes (7,8) Light, 1932
Haspidotermes (6) Miller, 1991
Hesperotermes (6) Gay, 1971
Homallotermes (4) John, 1925
Hoplotermes (8) Light, 1933
Labiocapritermes (4) Krishna, 1968
Labritermes (4) Holmgren, 1916
Lepidotermes (2) Sjöstedt, 1924
Lophotermes (6) Miller, 1991
Macrognathotermes (6) Miller, 1991
Malaysiocapritermes (4) Ahmad and Akhtar, 1981
Megagnathotermes (2) Silvestri, 1914
Microcerotermes (1–6, 8) Silvestri, 1901
Microcapritermes (4) Holmgren, 1914
Mucrotermes (2) Emerson, 1960
Okavangotermes (2) Coaton, 1971
Ophiotermes (2) Sjöstedt, 1924
Oriencapritermes (4) Ahmad and Akhtar, 1981
Orthoganthotermes (8) Holmgren, 1910
Orthotermes (2) Silvestri, 1914
Ovambotermes (2) Coaton, 1971
Paracapritermes (6) Hill, 1942
Pericapritermes (2, 4, 5) Silvestri, 1914
Pilotermes (2) Emerson, 1960
Planicapritermes (8) Emerson, 1949
Profastigitermes (2) Emerson, 1960
Prohamitermes (4) Holmgren, 1912
Promirotermes (2) Silvestri, 1914
Protocapritermes (6) Holmgren, 1912
Protohamitermes (4) Holmgren, 1912
Pseudhamitermes (4) Noirot, 1966
Pseudocapritermes (4) Ahmad and Akhtar, 1981
Pseudomicrotermes (2) Holmgren, 1912
Quasitermes (3) Emerson, 1950
Saxatilitermes (6) Miller, 1991
Syncapritermes (4) Ahmad and Akhtar, 1981
Synhamitermes (4, 8) Holmgren, 1912
Termes (2, 5–6, 8) Linné, 1758
Thoracotermes (2) Wasmann, 1911
Trapellitermes (2) Sands, 1995
Tuberculitermes (2) Holmgren, 1912
Unguitermes (2) Sjöstedt, 1924
Unicornitermes (2) Coaton, 1971
Xylochomitermes (6) Miller, 1991

Continued

Table 1 *Continued*

TERMITINAE *Continued*

Cornicapritermes (8) Emerson, 1950
Coxocapritermes (4) Ahmad and Akhtar, 1981
Crepititermes (8) Emerson, 1925
Eremotermes (1, 2, 4) Silvestri, 1811
Euchilotermes (2) Silvestri, 1914
Fastigitermes (2) Sjöstedt, 1924
Incolitermes (6) Gay, 1966
Inquilinitermes (8) Matthews, 1977
Kemneritermes (4) Ahmad and Akhtar, 1981
Neocapritermes (8) Holmgren, 1912
Nitiditermes (2) Emerson, 1960
Noditermes (2) Sjöstedt, 1924
Proboscitermes (2) Sjöstedt, 1924
Procapritermes (4) Holmgren, 1912
Procubitermes (2) Silvestri, 1914
Sinocapritermes (4) Ping and Xu, 1986
Spicotermes (8) Emerson, 1950
Spinitermes (8) Wasmann, 1897

NASUTITERMITINAE Hare, 1937

Aciculioiditermes (4) Ahmad, 1968
Acicultermes (4) Emerson, 1960
Afrosubulitermes (2) Emerson, 1960
Agnathotermes (8) Snyder, 1926
Ahmaditermes (4) Akhtar, 1975
Alstonitermes (4) Thakur, 1975
Ampoulitermes (4) Mathur and Thapa, 1962
Angularitermes (8) Emerson, 1925
Anhangatermes (8) Constantino, 1990
Araujotermes (8) Fontes, 1982
Arcotermes (4) Fan, 1983
Armitermes (8) Wasmann, 1897
Atlantitermes (8) Fontes, 1979
Australitermes (6) Emerson, 1960
Baucaliotermes (2) Sands, 1965
Bulbitermes (4) Emerson, 1949
Caetetermes (8) Fontes, 1981
Cahuallitermes (7) Constantino, 1994
Ceylonitermellus (4) Emerson, 1960
Ceylonitermes (4) Holmgren, 1912
Coarctotermes (3) Holmgren, 1912
Coatitermes (8) Fontes, 1982
Coendutermes (8) Fontes, 1985
Constrictotermes (8) Holmgren, 1910
Convexitermes (8) Holmgren, 1910
Cornitermes (8) Wasmann, 1897
Cortaritermes (8) Coaton, 1962
Cucurbitermes (4) Li and Ping, 1985
Curvitermes (8) Holmgren, 1912
Cyranotermes (8) Araujo, 1970
Cyrilliotermes (8) Fontes, 1985
Diversitermes (8) Holmgren, 1912
Eleanoritermes (4) Ahmad, 1968
Embiratermes (8) Fontes, 1985
Emersonitermes (4) Mathur and Sen Sarma, 1959
Enetotermes (2) Sands, 1995
Ereymatermes (8) Constantino, 1991
Eutermellus (2) Silvestri, 1912
Fulleritermes (2) Holmgren, 1912
Grallatotermes (2, 4, 5) Holmgren, 1912
Havilanditermes (4) Light, 1930
Hirtitermes (4) Holmgren, 1912
Hospitalitermes (4, 5) Holmgren, 1912
Ibitermes (8) Fontes, 1985
Kaudernitermes (3) Sands and Lamb, 1975
Labiotermes (8) Holmgren, 1912
Lacessititermes (4) Holmgren, 1912
Leptomyxotermes (2) Sands, 1965
Leucopitermes (4) Emerson, 1960
Longipeditermes (4) Emerson, 1960
Macrosubulitermes (6) Emerson, 1960
Macuxitermes (3) Cancello and Bandeira, 1992
Malagasiotermes (3) Ahmad, 1960
Malaysiotermes (4) Ahmad, 1968
Mimeutermes (2) Silvestri, 1914
Mycterotermes (2) Sands, 1965
Nasopilotermes (2) He, 1987
Nasutitermes (2–6, 8) Dudley, 1980
Obtusitermes (8) Snyder, 1924
Occasitermes (6) Holmgren, 1912
Occultitermes (6) Emerson, 1960
Oriensubulitermes (8) Emerson, 1949
Paracornitermes (8) Emerson, 1949
Parvitermes (8) Emerson, 1949
Periaciculitermes (4) Li, 1985
Peribulbitermes (4) Li, 1985
Postsubulitermes (2) Emerson, 1960
Proaciculitermes (4) Ahmad, 1968
Procornitermes (8) Emerson, 1949
Rhadinotermes (2) Sands, 1912
Rhynchotermes (8) Holmgren, 1912
Rotunditermes (8) Holmgren, 1910
Sinonasutitermes (4) Li and Ping, 1986
Spatulitermes (2) Coaton, 1971
Subulioiditermes (4) Ahmad, 1968
Subulitermes (8) Holmgren, 1910
Syntermes (8) Holmgren, 1910
Tarditermes (2) Emerson, 1960
Tenuirostritermes (7, Holmgren, 1912
Triangularitermes (8) Matthews, 1977
Trinervitermes (2, 4) Holmgren, 1912
Tumulitermes (6) Holmgren, 1912
Velocitermes (8) Holmgren, 1942
Verrucositermes (2) Emerson, 1960

Key to geographical regions: (1) Palearctic, (2) Ethiopian, (3) Malagasy, (4) Indomalayan, (5) Papuan, (6) Australian, (7) Nearctic, (8) Neotropical.

*c.*110 million years old, is *Meiatermes araipena,* found in limestone deposits. The first fossil soldier of the family Mastotermitidae from the upper Oligocene/lower Miocene was found in amber in Mexico by Krishna and Emerson (1983). Mastotermitidae have also been found in fossils from the Oligocene period in Europe (Nel and Arillo, 1995)

In the Eocene rocks, Mastotermitidae and Kalotermitidae are common, while in the Oligocene, in Baltic amber, Hodotermitidae, Kalotermitidae and Rhinotermitidae have been found.

RELATIONSHIP TO COCKROACHES

Termites are closely related taxonomically to wood-eating cockroach ancestors. The Australian termite present today, *Mastotermes darwiniensis,* has many similarities to cockroaches. It lays its eggs in batches of 16 to 24 held together in two rows, and also has a similar anal lobe on the hind-wing, and protozoa and bacteria in its gut. Considering common characteristics it is suggested that Blattodea and Cryptocercidae are a sister group of Mantodea while Isoptera is a sister group of this complex. As termites are closely linked to cockroach-like ancestors they may have initially lived inside rotten wood. Termites living inside wood could have burrowed into the ground, or tunnelling down into the roots of a plant or tree could have led to a subterranean existence. When the nests became large, or all the food supply had simply run out, the termites were compelled to search elsewhere. Where subterranean conditions were not always suitable, it is possible that some termites then started to build mounds. Nest separation also led to stability as the nest was not the food source. Small colonies within wood, which do not need to regulate temperature, most likely gave rise to larger colonies in soil where temperature regulation was developed.

Abe (1987) has looked at the evolution of nest and feeding habits based on the kind of food source wood used. He speculates on evolutionary trends related to problems, advantages and distribution of termites living in a one piece type nest as in Kalotermitidae, intermediate type as in Rhinotermitidae and separate as in Termitidae. He speculates that the separate type arose from the intermediate, but lost the ability for 'rafting' (common in the intermediate types) so that fungus-growing termites could not reach Australia and the Americas.

To survive in dry conditions primitive wood-dwelling termites developed elaborate rectal pads for absorbing water from the faeces before they are released. Higher termites, including the soil dwellers, do not possess such elaborate rectal pads but are able to burrow down to the water table for water. Mastotermes and Termopsidae families contain the most primitive termites. However, the alates (winged reproductives) of these and other primitive termites tend to be the stronger fliers. Primitive termites also tend to possess protozoa in their gut. A reduction of the number of sternal

Table 2. Geological time scale showing the presence of termite families.

Millions of years (approximate)		Era	Period	Events (approximate time scale)
3500		Azoic		'Life beings'
2000		Pterozoic	Precambrian	
				First animal fossils
570				
			Cambrian	Sea life and early fish
500		P		
		a	Ordovician	
440		l		
		a	Silurian	First land plants
395		e		
		o	Devonian	
345		z		Early amphibians and reptiles
		o	Carboniferous	Large forests
280		i		First cockroaches
		c	Permian	First termites possibly
225				
		M		
		e	Triassic	Early mammals
195		s		First wasp-like ancestor
		o	Jurassic	Early birds
136		z		
		i	Cretaceous	Earliest termite fossils:
		c		Hodotermitidae and
65				Mastotermitidae
			EPOCHS	
			Palaeocene	
54		T		Mastotermitidae and
		e	Eocene	Kalotermitdae fossils common
38	C	r		
	e	t	Oligocene	Rhinotermitidae fossils common
26	n	i		
	o	a	Miocene	Termitidae fossils common
7	z	r		
	o	y	Pliocene	
2	i			First humans
	c	Quarternary	Pleistocene	
Recent			Holocene	

glands is also found in the more advanced termites such as the Termitidae (Noirot, 1995). Fossils suggest that workers probably evolved after the presence of winged forms. Once workers existed, then termites became social insects. Drywood termites have alates, soldiers and nymphs that act as workers in the colony but can change to alates. The origin of fungus growing may have come from carton (thin paper from wood) or faeces in nests which became infected with fungus.

CASTES OF TERMITES

An excellent introduction to the sociality of termites and the relationship between different castes is given in E.O. Wilson's *The Insect Societies* (1971). A termite colony consists of workers or pseudoworkers, soldiers, reproductives, larvae and eggs (Plates 1–3).

Workers

Workers are wingless, not sexually mature and, apart from the family Hodotermitidae, are blind. As with other social insects, termite workers have an important role in the nest. Their tunnelling and food collection makes some of them pests. Other jobs include building and maintaining the nest, looking after and feeding the young and other non-wood feeding castes such as soldiers and the royal pair. Workers also groom and clean other castes. Examples of task-specific activities for two worker castes have been examined by Lys and Leuthold (1991).

Termites known as the lower termites are considered the most primitive because they have protozoa in their hind gut that help to break down cellulose to sugars which then become available to the termite. In this group, consisting of families Mastotermitidae, Rhinotermitidae and Termopsidae, the workers are often termed pseudoworkers or pseudergates, whereas in the Kalotermitidae they are referred to as nymphs. These can develop as larval stages in the normal way or from regressive moults (losing older stage characteristics). Often their specific roles in the colony are not obvious, as they may carry out many different tasks. However, some different roles may be detected, as seen in *Zootermopsis* where 1st and 7th instars have been shown to be the most active. Colonies are also plastic, in that these false workers can change into a different caste where required.

In the other group of termites, which are called the higher termites, worker termites have different roles and are often of two or more sizes or castes. Size may also be linked to sex. An example is seen in *Macrotermes michaelseni* where minor workers are female, and major workers are male. *Macrotermes* workers have been shown to be separated into two age groups which perform different functions within the colony (Badertscher *et al.*, 1983). Between 5 and 30 days old they forage for plant material, feed larvae and feed on the fungus found on combs in their nest. Those greater than 30 days old carry food and eat the actual fungus comb to produce faeces. In *Hodotermes mossambicus* most duties are carried out by 6th instar larvae, while older workers forage in the open and the young, unpigmented workers feed on collected food in the nests. In *Nasutitermes,* the oldest workers are involved with nest repair and foraging.

As well as changing roles at different ages, each sex may have different duties. In *Odontotermes* the queen is attended by female workers while the foraging is done by male workers, as in *Macrotermes.*

Soldiers

The role of soldiers is for defence. They possess a larger head that is longer and wider than that of the workers so as to contain more muscle or an enlarged frontal gland. Size may also be related to sex or different developmental pathways in some species. Often the soldier head is coloured yellow to brown with enlarged mandibles of various shapes or sizes. Soldiers that have enlarged jaws, or reduced mandibles, cannot feed themselves and they have to rely on workers for this.

Differing sizes of soldiers are found in some species. Major and minor soldiers are common in some Termitinae. In the Rhinotermitinae *Schedorhinotermes* has two, or in some species, three kinds of soldier that differ in size, headshape and mandibles. Three sizes of soldiers are also found in *Acanthotermes acanthothorax* (Macrotermitinae). Some termites may have no soldiers at all (soldierless).

In some species the mandibles are reduced and the head is modified into a long nose from which a toxic glandular secretion is fired from the tip. Figures 2 and 3 show some of the commonest forms of soldiers and examples of the variety of forms that can exist.

In some termite genera soldiers may be different sexes. *Amitermes evuncifer* can have equal proportions of males and females. Macrotermitinae are mainly all females and Nasutitermitinae are males. In some cases one may find soldier/reproductive intercastes where some of the female characteristics, for example, wing rudiments or ovaries with eggs, may be found in the soldier. Intercastes are often found in the genus *Zootermopsis.* These are common when the original reproductives are missing.

Reproductives

Alates or swarmers

These are the winged reproductives which can produce new kings and queens.

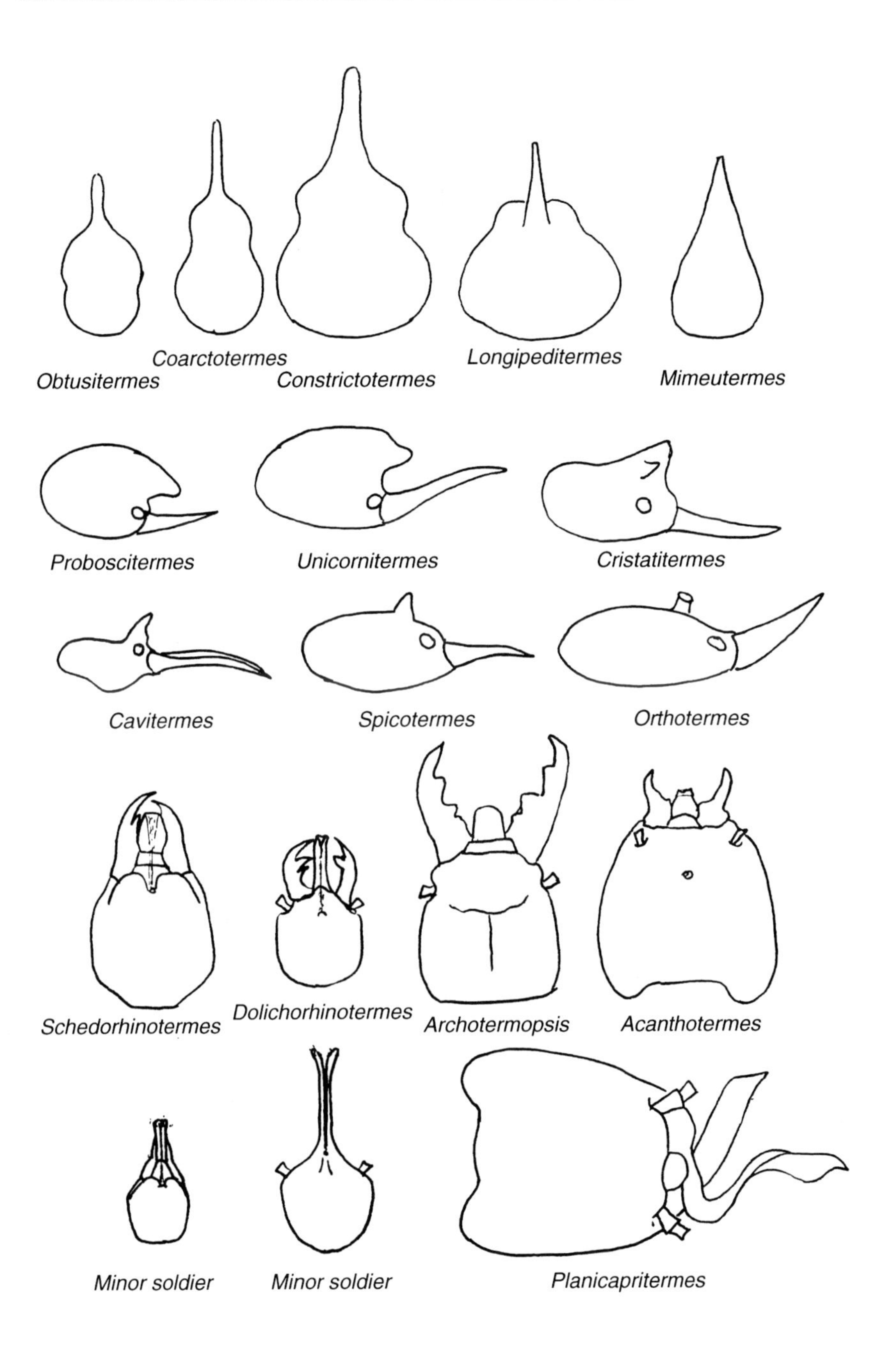

Fig. 2. Examples of head shape variation in soldier termites.

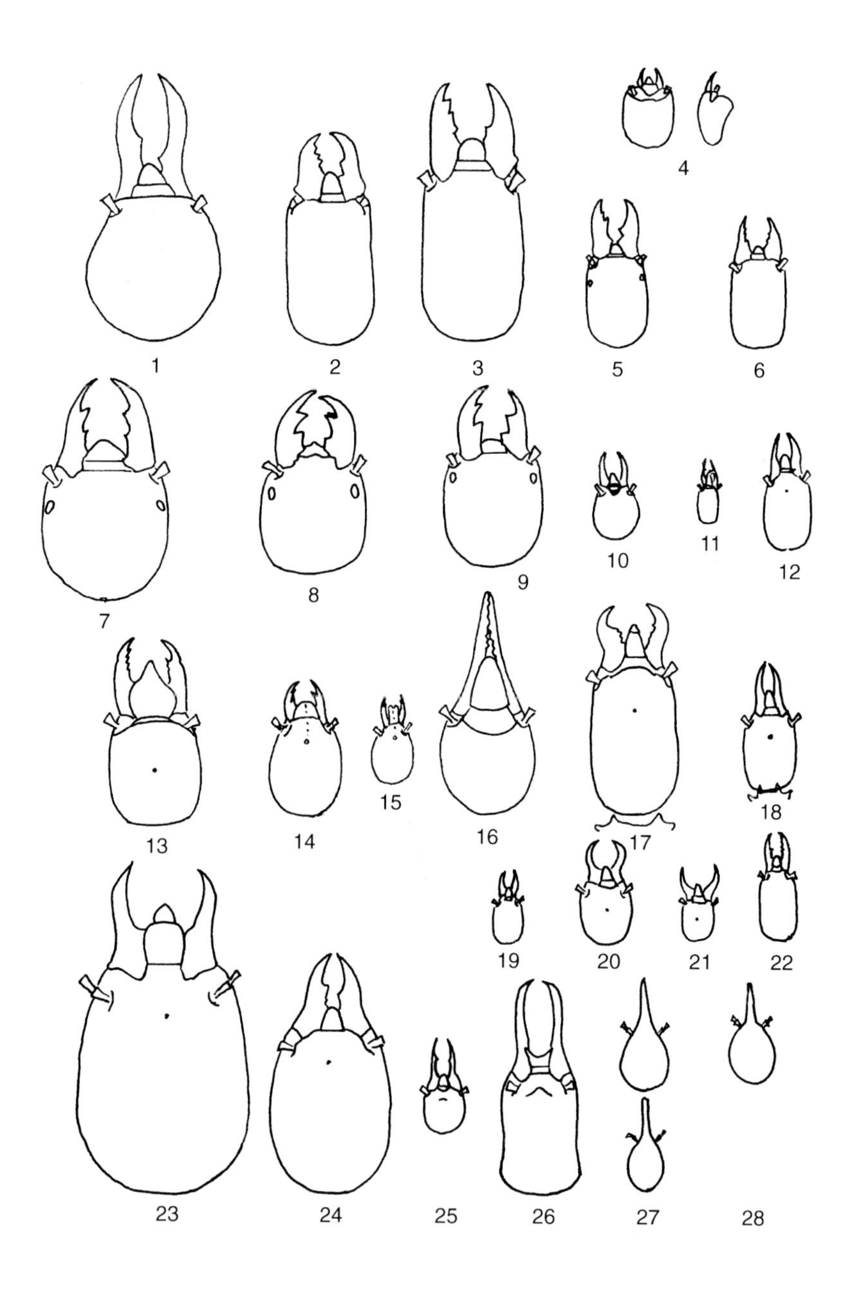

Fig. 3. Termite soldier heads and the main distinguishing features.

Key to soldiers.

Number	Genus	Main distinguishing features
1	*Mastotermes*	Head round, five segmented tarsi, Australian
2	*Neotermes*	Large teeth, kidney shaped pronotum
3	*Kalotermes*	Large teeth, head flattened towards the front
4	*Cryptotermes*	Head with flat front, small curved mandibles
5	*Bidfiditermes*	Large teeth, pronotum with straight sides
6	*Glyptotermes*	Head narrower than *Bifiditermes* and mandibles shorter
7	*Hodotermes*	Eyes present, lower base of mandibles reddish
8	*Microhodotermes*	Eyes present, mandibles black, small lump on second tooth of right mandible (absent in *Hodotermes*)
9	*Anacanthotermes*	Eyes present, two large teeth on right mandible, found in North Africa and Sahel
10	*Coptotermes*	Rounded head narrower at front with large opening
11	*Heterotermes*	Labrum with point on the tip
12	*Retculitermes*	Similar appearance to *Heterotermes* but point on labral tip absent
13	*Psammotermes*	Head squarish, labrum inflated
14	*Schedorhinotermes* (major soldier)	Groove from fontanelle to end of labrum
15	*Schedorhinotermes* (minor soldier)	As for major soldier but labrum long
16	*Serritermes*	Mandibles straight and teeth all along the length
17	*Pseudacanthotermes* (Major soldier)	Two points on anterior pronotum
18	*Pseudacanthotermes* (Minor soldier)	As for major soldier
19	*Microtermes*	Very small termites, straight mandibles, antennae 12–14 segments
20	*Ancistrotermes* (Major soldier)	Size as for Microtermes, mandibles curved upwards at tip, antennae 14–15 segments
21	*Ancistrotermes* (Minor soldier)	As for major soldier
22	*Microcerotermes*	Elongated head, mandibles with saw-like teeth
23	*Macrotermes*	Large termites, white triangular tip to labrum, no teeth found midway on mandibles
24	*Odontotermes*	Size may be similar to minor workers of *Macrotermes*. Single tooth midway on right mandible or on both mandibles
25	*Eremotermes*	Long mandibles with curved end, single tooth found mid-way on each mandible
26	*Cubitermes*	Head rectangular slightly curved inwards at sides, labrum two points at the end
27	*Trinervitermes*	Long nose, major and minor soldiers are found which have no mandibles
28	*Nasutitermes*	One size of soldier, only mandibles present but reduced to small points

Males can be distinguished from females by the presence of styles on the 9th sternal segment. The female also has an enlarged and extended 7th sternal segment and split 8th and 9th segments (Fig. 4).

A certain size of colony is needed before alates are produced. In *Cubitermes fungifaber* this is approximately 6000 individuals. It often, therefore, takes several years before a colony reaches a size where alates are produced, e.g. eight years for *Coptotermes formosanus* in China. Alates leave the colony *en masse* and alate flights contain a few hundred individuals for lower termites or many thousands for higher termites.

Before flight, alates congregate away from the main colony. They then leave from holes or slits in the ground, mound or wood or from special flight turrets. These are built or constructed at fixed times of day or night depending on the species of termite and environmental conditions. Slits that are made causing damage to living wood can be repaired by the tree with new tissue (callus). The number of exit holes can vary from a few to several hundred, especially in the mound builders. Before emergence, workers prepare the exit holes or build the flight turrets while soldiers stand guard. A waiting room is often constructed close to the site of release or even at the base of a turret. Here the alates congregate so that a quick escape is possible. Rainfall may be the trigger for initial release and for other flights occurring later when there is no rain. Different sexes emerge together, but in some species there may be more males present ensuring that each female will find a mate.

In lower termites, and in temperate regions, flights may occur over several months or several times a year. With higher termites, and especially in dry regions after rain, there is a definite time of year and day for flights and these are synchronized with other flights in the same region. Most termites

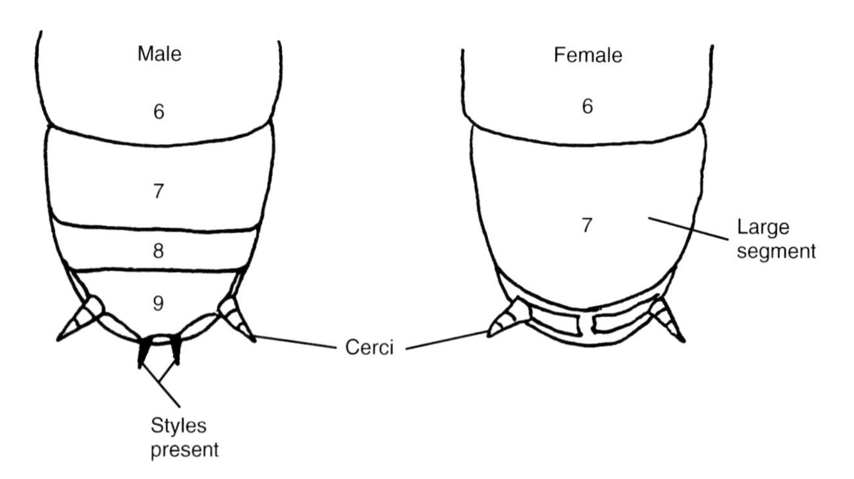

Fig. 4. Ventral view of the abdomen of male and female alates showing differences.

are poor fliers and therefore do not fly great distances. Lower termites fly better than higher ones. *Coptotermes* can fly for a few hundred metres, especially when aided by strong wind. The wind can determine the direction of flight but greater dispersal may mean there is less chance of finding a mate.

When alates land, they lose their wings immediately or following contact with the opposite sex (Plate 4). This is achieved by raising and twisting the abdomen, or by rubbing against an obstacle which breaks each wing off at the weak sutures. One genus, *Pseudacanthotermes,* flies with the male grasping hold of the back of the female. Some alates, such as *Allodontermes giffardi* do not fly, but instead climb up grasses to form tandems.

Alates have sternal and tergal glands on their abdomens (Ampion and Quennedey, 1981). The tergal glands of the female produce a volatile compound responsible for attracting males. The sternal glands of both sexes are used to lay down a scent trail so that contact is not lost between the pair as they travel in tandem looking for a place to burrow. Once the members of a pair of alates meet there is no time for extensive courtship and mating as they must dig out a tunnel to avoid predators and other hazards. Soil dampened by rain allows alates of soil-nesting species to burrow quickly, with both sexes taking part. Wood-dwelling termites look for crevices and cracks in wood which they enter and enlarge. Mating may not take place straight away but after a few weeks.

Alates from some fungus-growing termites can ingest fungus comb before flight. The fungus is carried as a bolus in the rectal chamber and is used by the first workers to inoculate the new comb built. Other fungus-growing termites may depend on other sources of spores from fungi growing on the soil surface which the first workers collect. The presence of these fungi and the presence of new foraging workers has to coincide for this method to be successful.

Millions of alates may be killed by predators, especially ants and birds, on release and even if they manage to pair and burrow they may still not survive. Large colonies of Macrotermitinae can experience as much as 50% mortality; in the Rhinotermitidae, it may be even higher.

Royal pair

After burrowing into soil, or wood in the case of timber-living species, the entrance is sealed. In the drywood termite, a lattice is first constructed across the opening using the mouthparts, and this is then filled in with an abdominal secretion. Once under the ground or inside wood, a copularium is constructed. The alate pairs, which have burrowed into soil, do not feed but rely on their fat store and protein breakdown from wing muscles. Some may, however, eat some of their first born.

The abdomen of the female (queen) becomes swollen in many termites (Fig. 1, 5). The time taken for large changes in size can be eight months after colony foundation for *Macrotermes subhyalinus* female alates. *Macrotermes* and *Odontotermes* queens can become thumb-sized egg-laying machines. Circulation and respiration is helped by peristaltic movements of the body. The trachea are wider and thicker than in the other castes so as to provide an adequate air supply needed for energy requirements. The king, however, only increases slightly in size and may be only a fifth of the length of the queen with an abdomen a hundred times smaller. The more primitive termite queens living inside wood produce only a few eggs per day. A *Coptotermes* queen can produce 100 eggs per day but more advanced termite queens, such as *Macrotermes,* can produce 30,000–40,000 eggs per day and up to 10 million a year. Fertilization of eggs may, in some species, be monthly. *Odontotermes obesus* queens can produce one egg every few seconds. Eggs may be kept in specific areas or in chambers within the nest. In fungus growers they may be placed near the fungus combs which provide a moist atmosphere (Plate 3). Some eggs may be eaten, providing an extra food supply. In other species they may be placed near chewed wood or frass (seed-like faecal pellets), as is the case of drywood termites, which may help in maintaining a humid environment. The king and queen feed the newly hatched young in the first brood from body food stores, in the case of soil-living termites, and from food derived from surrounding wood in the case of wood-dwelling termites.

Eventually the flight and mandibular muscles of the queen degenerate, as do the eyes. The queen is attended by many workers who groom her all over, remove eggs and feed her. During cleaning of her body and when removing eggs workers receive secretions, which contain chemical messages as well as being attractive. Primary queens can survive for 15–20 years and some may live for 50. Workers, however, may only live for a few months. *Mastotermes* kings have been recorded as living for 16 years.

In some species many queens may be found in the same nest. All queens are found in definite cells or chambers. In the Macrotermitinae this queen cell is a hard case with only a few exit/entry holes in rows around the side. Figure 5 shows an opened cell with several queens present. The hard cell protects the queen and makes it easier to defend against intruders. The queen can pull herself along using her legs and contraction of her body. Although it appears that she is trapped for life in this cell, she can, with the help of workers, be moved out of the chamber when under threat of attack.

Replacement reproductives

If a queen dies or is getting old, and her egg production is decreasing, substitute queens may be produced. These are commonly found in lower termites and also can exist in some of the higher termites such as *Cubitermes, Macrotermes, Nasutitermes* and some *Microcerotermes.* Different species of termite (e.g. in the genus *Nasutitermes)* may vary in the amount of reproductive replacement and formation (budding) of new colonies. Replacement reproductives are not only important for the continued survival of a colony but also can allow for formation of closely related satellite colonies, separate from the main colony. These are very mobile colonies which, as well as helping with survival in adverse conditions, have major pest implications.

Supplementary reproductives with wing pads (brachypterous neotenics) are very common. Most supplementaries are darker than normal workers and are often deep yellow to brown in colour (Plate 1). The duration for replacement can vary depending on the availability of possible substitutes, the size of the colony and many other factors. In drywood termites this can be a few days, in *Mastotermes* around a month, while in some *Nasutitermes* it may take six weeks. In *Cryptotermes,* a new colony can be produced from just four individuals. Several supplementaries may be produced but only the fittest survive, the others being eaten. However, one can often find a few supplementaries together that are egg laying, especially in lower termites. Supplementaries often have a shorter life than true reproductives and their egg-laying capacity is lower, but with many supplementaries present the overall numbers of termites produced can be large. This is the reason why numbers of *Reticulitermes* can reach millions in a short time, which would not be possible with colonies derived from alates only (Thorne, 1996). Other kinds of supplementary reproductives (ergatoid or apterous neotenics) can be produced from wingless workers. Replacement

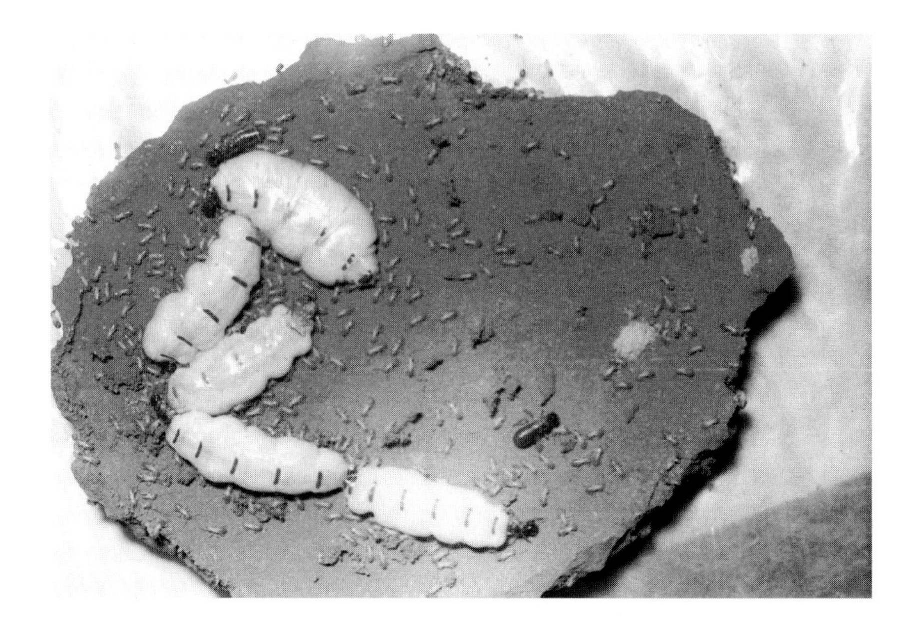

Fig. 5. Five queens and two kings of *Macrotermes* in an opened royal cell.

reproductives may, as with the royal pair, spend most of their time involved in receiving or giving food to others (trophallaxis), rather than performing other worker duties. This is common in *Zootermopsis*.

Figure 6 summarizes the pathways for caste production in a lower termite and a higher termite. The example in lower termites is for the family Kalotermitidae, where the workers can develop into soldiers or nymphs which are destined to become alates. In the other lower termites, such as the Rhinotermitidae, pseudergates are thought to remain immature for their entire life and can change to soldiers and alates when required. The chemical messages which regulate these changes are distributed by termites passing or receiving secretions from other termites and also possibly by volatile secretions (pheromones). Juvenile hormone (JH) is responsible for caste changes within the colony. In lower termites, the amount of JH is lowered at a certain time of year and alates are produced. The level of JH at moult-

Kalotermes flavicollis (Lower termite)

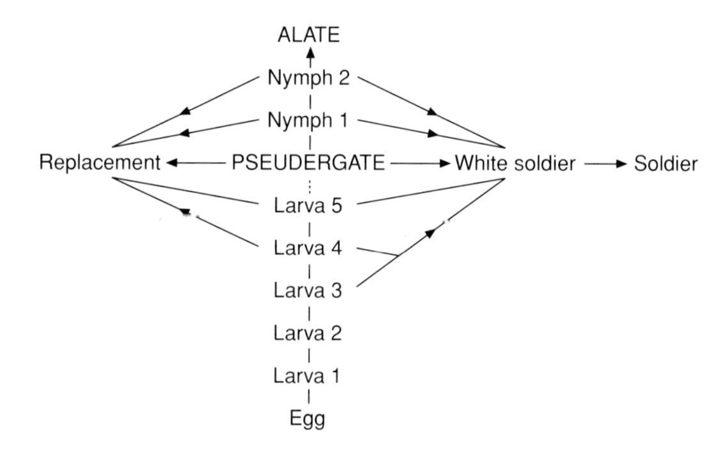

Macrotermes michaelseni (Higher termite)

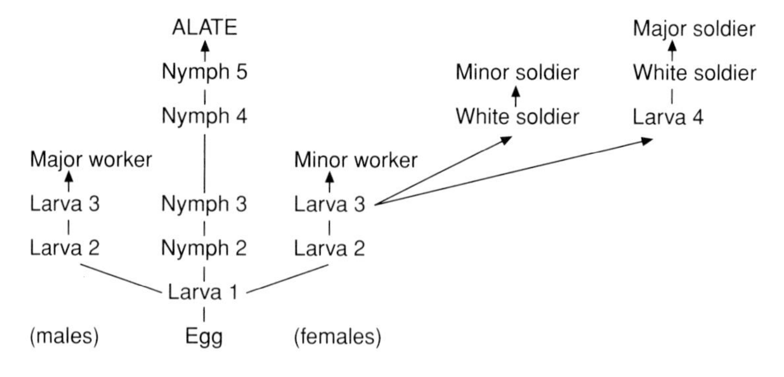

Fig. 6. Caste development in a lower and higher termite.

ing is also important for determining the change from workers to soldiers. High levels induce moulting to presoldier then soldier. The presence of a certain number of soldiers can inhibit the production of more soldiers. In higher termites low doses of JH cause worker development while high doses encourage soldier development (Lüscher, 1972; de Wilde and Beetsma, 1982). However, the developmental pathway may not be the same for all species in the same group. Australian Kalotermitidae are different from the European ones, which illustrates the plasticity of the group.

In lower termites the anal fluid of the queen was shown to inhibit the development of pseudergates into supplementary reproductives if a male is present. It is suggested that pheromones from a reproductive, act on the endocrine system stimulating the corpora allata to regulate JH level. In the absence of a queen, the king may possibly produce a pheromone which accelerates development of a queen. Production of different types of soldiers (major and minor) appears to be related to season in some termites. In *Cubitermes* the number of soldiers is a balance between the stimulatory influence of the reproductives and the inhibitory effect of the soldiers already present. JH content of *Macrotermes* eggs may also determine differentiation into neuter, larvae or reproductive. In the Termitidae soldiers can develop from large or small workers, male, female or both.

CLASSIFICATION OF TERMITES

Termites can be separated taxonomically using many different features: external morphology, internal features, food and nest type, chemical and behavioural differences.

External Morphology

Soldiers play an important part in termite classification. The most obvious differences are the shape and size of the head and mandibles (Plate 1). Figure 3 shows some of the more commonly found termites and also illustrates some of the extensive variations that can be found. Identification of soldiers can be made using mandible shape and arrangement, and the number and position of teeth. This can also be used for worker and imago mandibles but one may only be able to distinguish different genera rather than individual species using this feature. A primitive feature seen on the mandibles of some workers and alates is the presence of a subsidiary tooth between the apical and first marginal tooth. This is reduced or absent in more advanced termites and the number of marginal teeth increases and the apical tooth becomes more prominent. Plant feeders may have a deeply ridged molar surface on their mandibles whereas with humivorous feeders this is smooth or hollowed out.

If alates are collected and paired the first soldiers produced may not be typical of those in further generations. In *Mastotermes* the first soldier has no teeth. In the Rhinotermitidae and Termitidae a fontanelle (small hole in the head) exists for the release of a secretion from the frontal gland. The genus *Coptotermes* can be distinguished from *Globitermes* by the production of a white secretion from the frontal gland, the latter producing a yellow secretion. Compound eyes are present in *Mastotermes, Hodotermes* and *Anacanthotermes.* These are smaller in Termopsidae and absent in many other termites.

An unusual characteristic is found in the genus *Tetitermes,* where the swollen foretibia has an anterior spoon-like concavity that is used for digging. The postclypeus is sometimes a useful feature. French and American populations of *Reticulitermes* can be separated using this feature.

The number of segments in the cerci which are found either side of the end of the abdomen can vary from 2 to 3 in higher Termitidae to 5 in *Zootermopsis* and *Mastotermes*. The number of antennal segments may also be useful. *Microtermes* imagos, workers and soldiers have 15–16, 12–14 and 14 antennal segments respectively. *Ancistrotermes,* which may sometimes be hard to distinguish from *Microtermes,* has 17, 15 and 15 for imago, soldier and worker castes respectively. Wing microsculpturing is useful in distinguishing between genera and species, but less helpful for families and subfamilies. At least ten basic types of wing microsculptures have been described. The shape of papillae and micrasters on the wings are used, which can be simple structures or elaborate stars with 5–8 arms. Primitive termites have fewer types of microstructures on their wings. Figures 3 and 8–15 summarize some of the characteristics that can be used to distinguish soldiers, alates and workers of different termite families and subfamilies. Figure 7 illustrates standard measurements used for termites.

Internal Features

The form of the gut (Figs 15, 28) can be used for identification (Noirot and Noirot-Timothee, 1969; Johnson, 1979). In the Termitidae there is often a mixed segment, the malpighian tubules being on one side and proctodeal tissue on the other. In *Cubitermes* different species can be distinguished by the shapes of the diverticula. In soil-feeding termites the colon can be separated into two regions, one half containing filaments to which bacteria can be attached. The mixed segment is useful for identification of soldierless termites. Cuticular structures in the oesophagus, crop, proctodeal segment and enteric valves can also be used, as can size, shape, number and arrangement of spines in the gut. The gizzards of wood- and grass-feeding species have well-armed walls for this diet, as well as modified teeth and strong denticles. Denticles are absent in fungus growers and gizzard armature is also reduced in soil and root feeders.

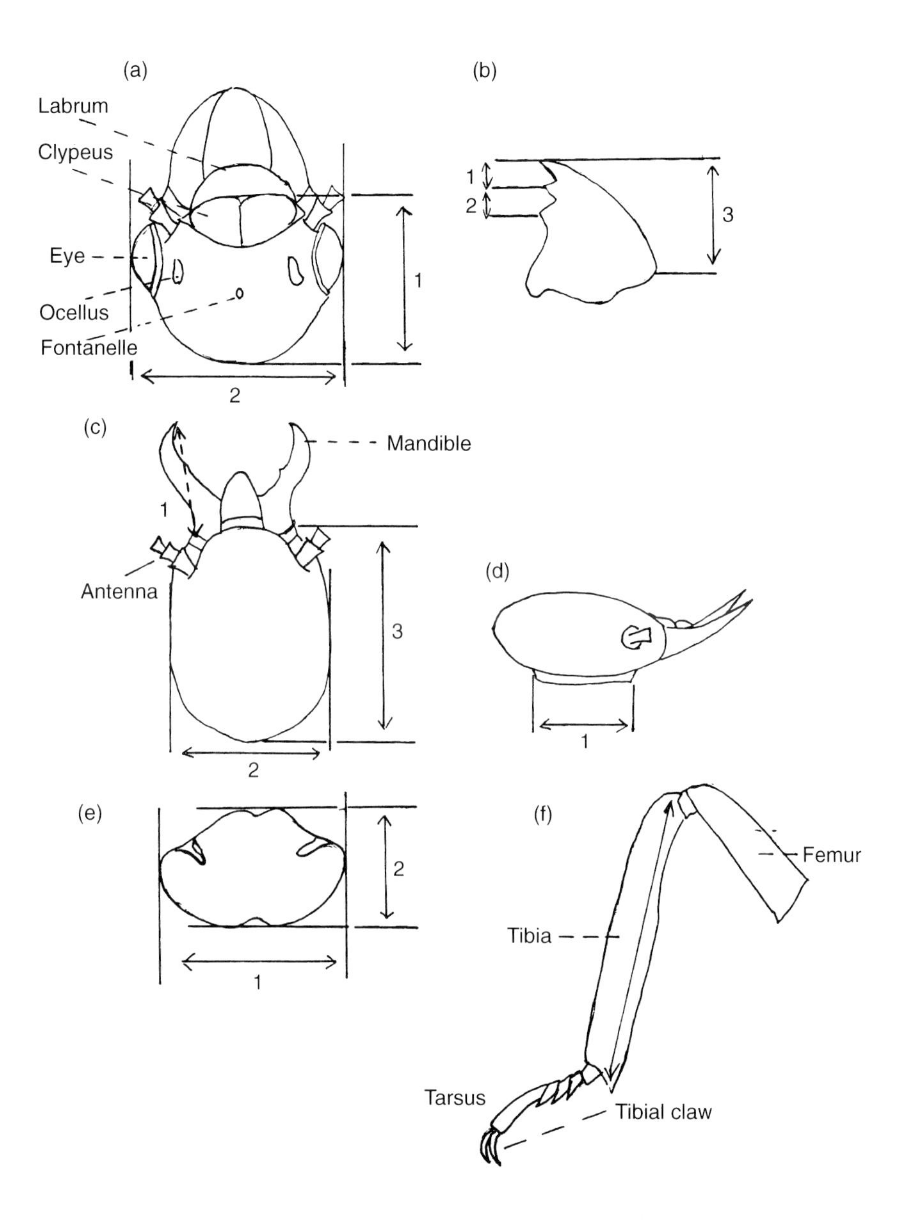

Fig. 7. Taxonomic measurements of termites. (a) Head of an imago (alate): 1, length; 2, width. (b) Mandible of imago: 1, distance of apical tooth to first marginal tooth; 2, distance of first to second marginal. (c) Soldier head: 1, length of left mandible; 2, width of head; 3, length of head. (d) Length of postmentum. (e) 1, width of pronotum; 2, length of pronotum. (f) Length of tibia.

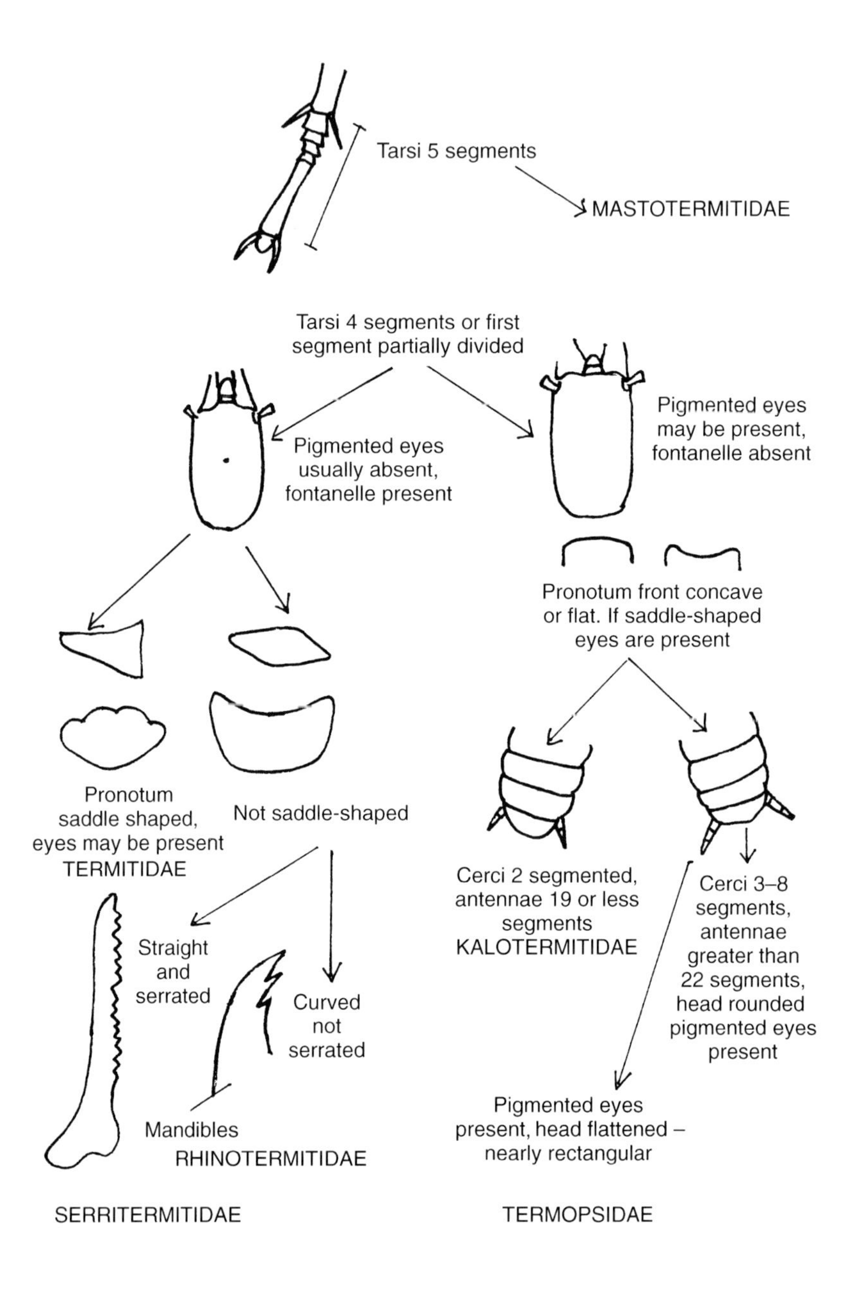

Fig. 8. Characteristics of termite families based on soldier characters.

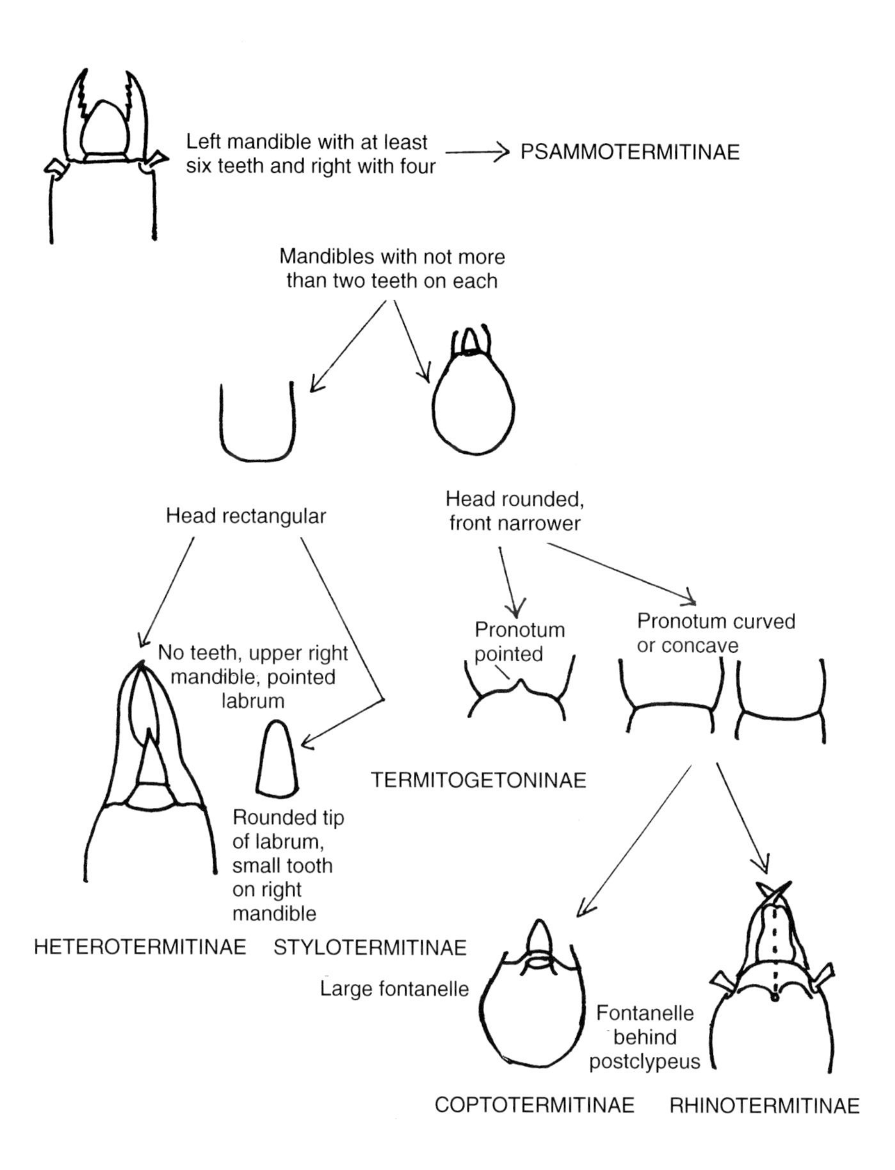

Fig. 9. General features of soldiers of sub-families of the family Rhinotermitidae.

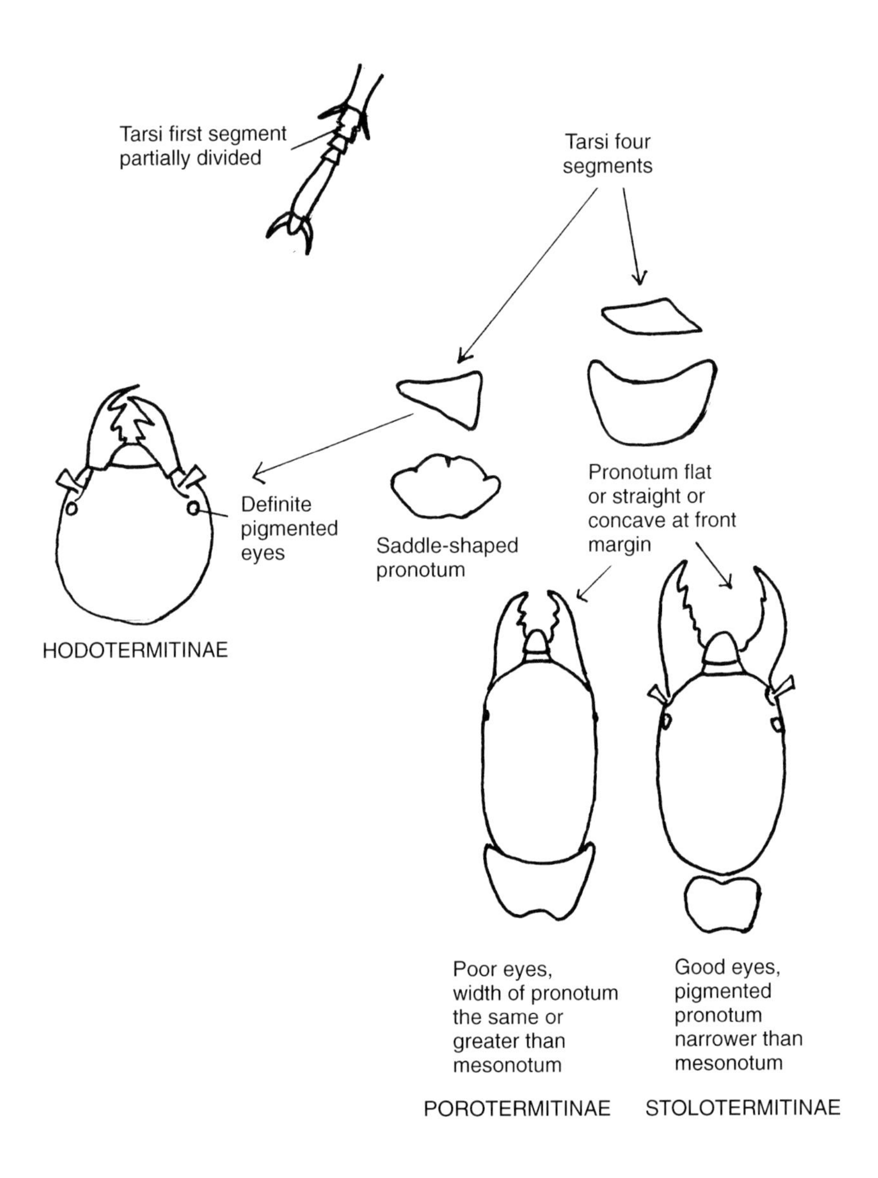

Fig. 10. General features of soldiers of sub-families of the family Hodotermitidae.

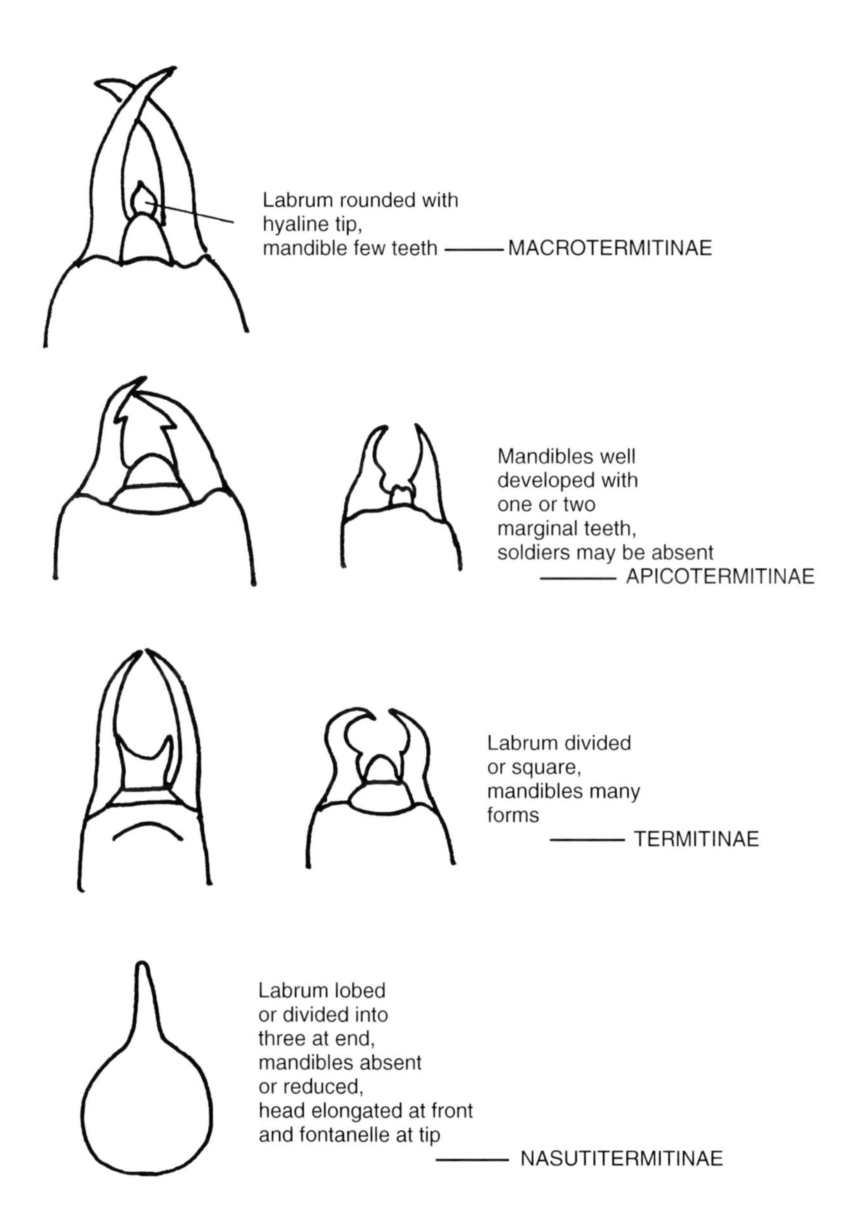

Fig. 11. General features of soldiers of sub-families of the family Termitidae.

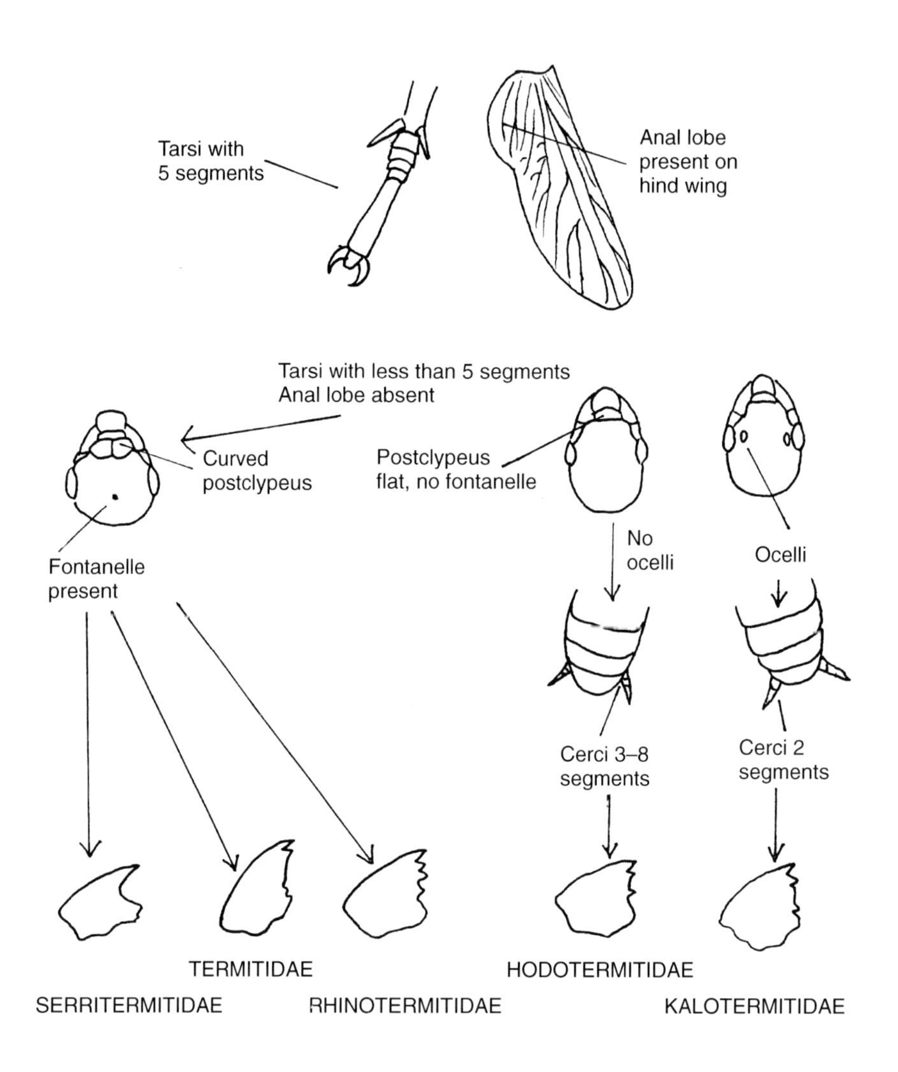

Fig. 12. Key to the termite families of imago/worker castes.

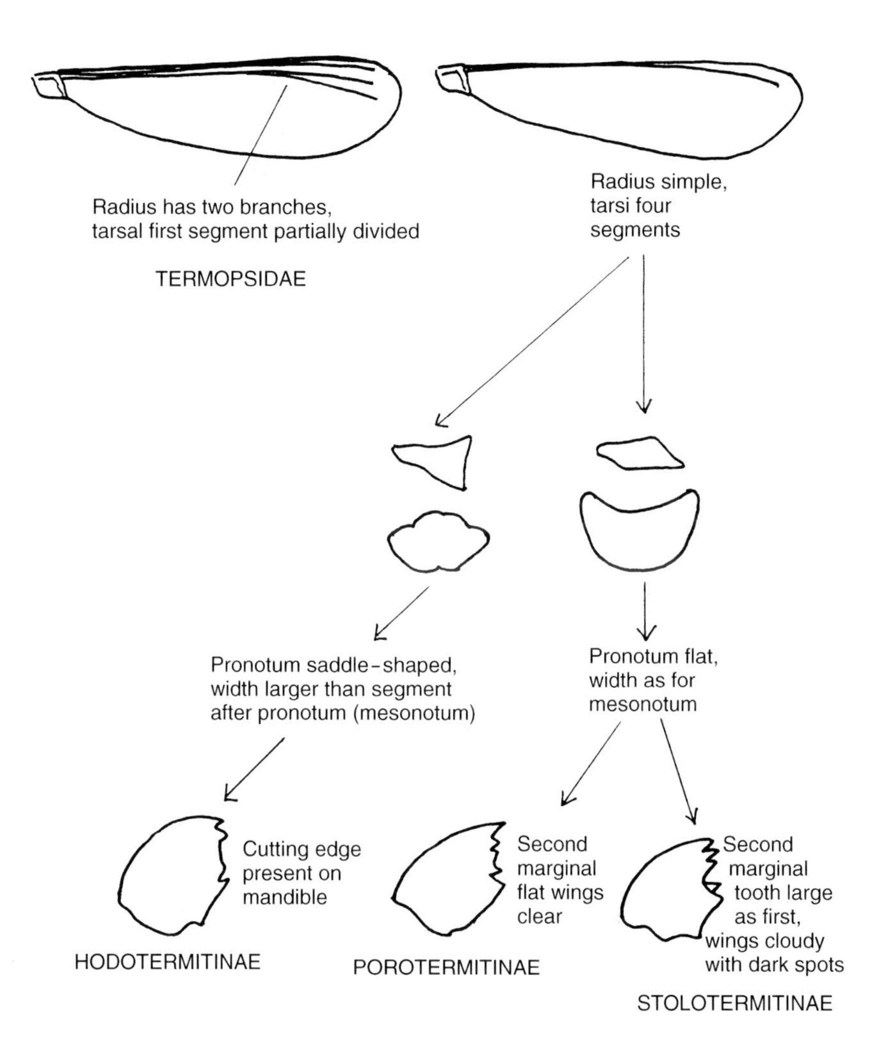

Fig. 13. General features of imago/workers in the sub-families of the family Hodotermitidae.

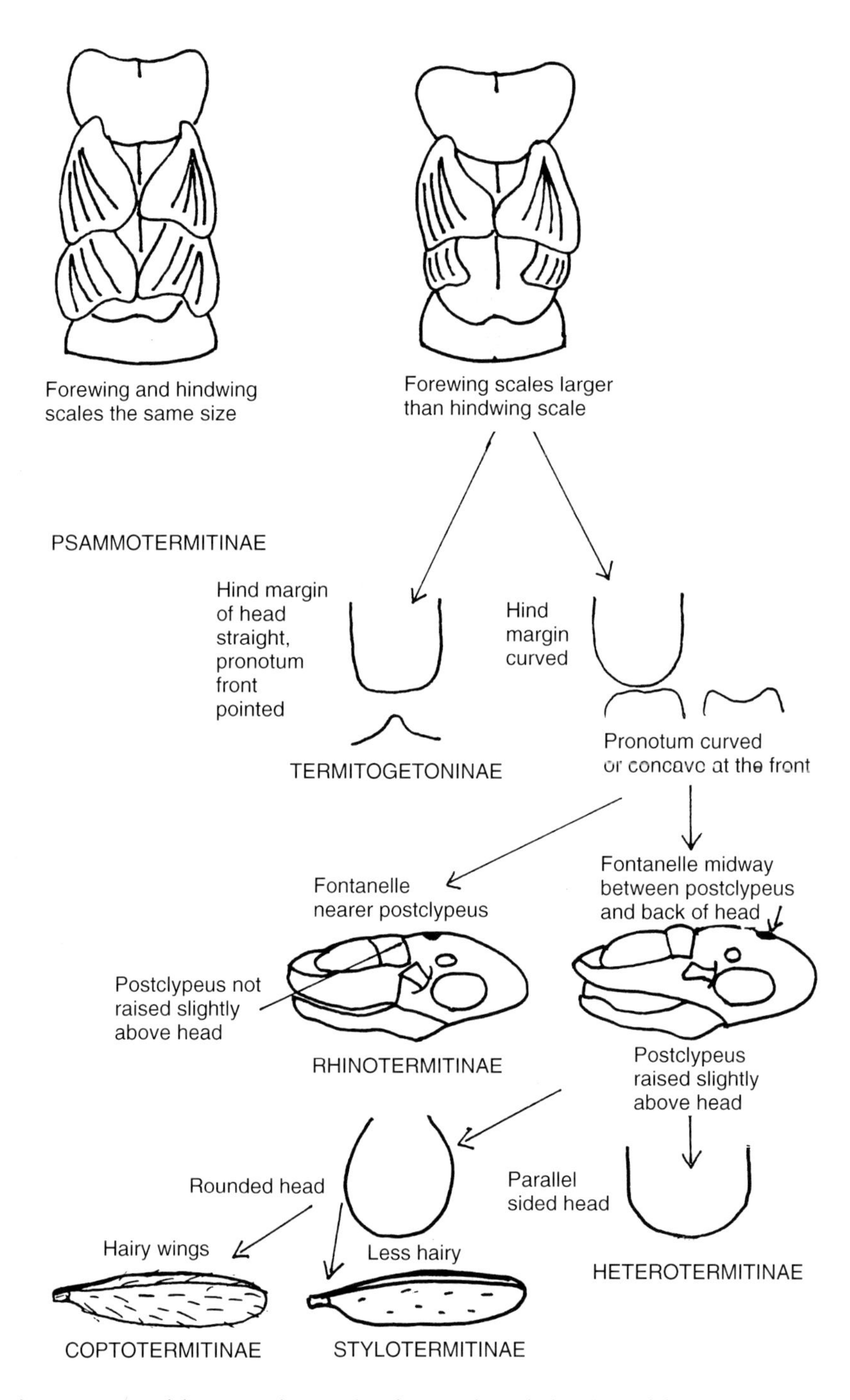

Fig. 14. General features of imago/workers in the sub-families of the family Rhinotermitidae.

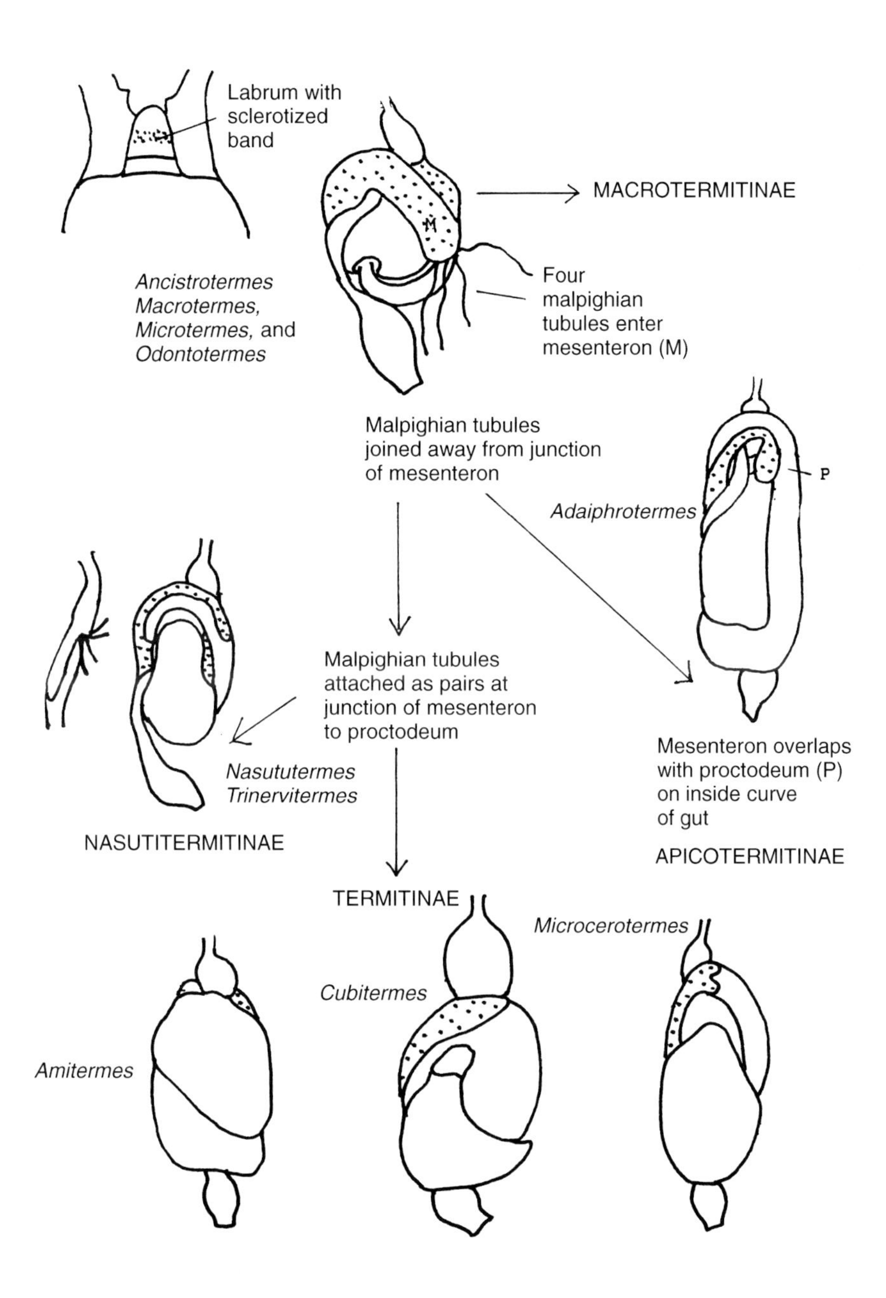

Fig. 15. General features of imago/workers in the sub-families of the family Termitidae (after Johnson, 1979).

Feeding Types

Termites can be divided into wood- and/or litter-feeding, fungus-growing, humus- and soil-feeding types, but some overlap often occurs. Worker mandibles and gut morphology related to food type can also be distinguished. The mandibles of grass and wood feeders have a molar plate, while soil feeders have a cusp used for crushing.

The presence of symbionts (Figs 16, 28) and fungus combs are also characteristic. Lower termites have protozoa and bacteria in their gut while higher termites have lost the protozoa and have only bacteria. Bacteria may actually be fixed to the cuticular lining of the hind gut of soil feeders such as *Cubitermes*. The frass (faecal pellets) of termites can be used for identification. In drywood termites the size and shape of pellets are important.

Nest Type

In some genera of termites there may be wide variation in nest type, e.g. *Odontotermes* (Akhtar and Anwer, 1991), and other distinguishing factors are needed to separate species within the genus. One method used has been the classification according to their habitat or nest type. Depending on where their nests are found they are also divided into drywood, dampwood and subterranean. Drywood termites are always found inside wood but dampwood and subterranean termites can also be found in the soil. The shape and position of nests can vary from species to species and also for the same species in different habitats. The micromorphology of nest material can also be used for identification purposes and will be described in more detail in Chapter 4.

Chemical Features

Some termites have a characteristic pH. Lower termites tend to have a lower total gut pH, while Termitidae have a higher pH. Cuticular hydrocarbon studies and DNA analysis have also been used to distinguish a few termite species and to look at relationships between genera and families. Soldier defence secretions also can be used.

Behavioural Features

Different species of the same genus can have different flight times which prevents interbreeding. This can be used to distinguish them (Wood, 1981). Often some species are night fliers, while others are day fliers.

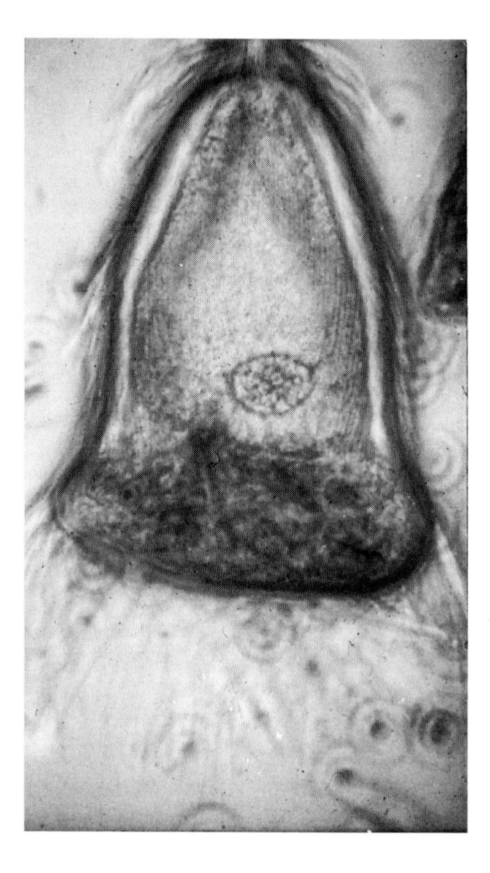

Fig. 16. The flagellate protozoa *Trichonympha* sp. from the gut of the dampwood termite *Zootermopsis*.

Night fliers have characteristically larger eyes and ocelli to gather as much light as possible. Day fliers may also be more darkly pigmented than the night fliers.

The study of aggressive behaviour may be useful for distinguishing different species or different colonies of the same species (Pearce *et al.*, 1990; Thorne and Haverty, 1991). Termite behaviour is discussed in more detail in Chapter 3.

Chapter 2

Distribution

WORLD DISTRIBUTION

There are over 2500 species of termites known today. Figure 17 illustrates the extent of termite distribution in the world. Table 1 gives a list of geographical distributions for different termite genera.

Termite numbers, species and nest variety increase as one moves towards the equator. Termite distribution can be related to temperature and rainfall. These change with latitude, and the limits of survival are between latitudes 45 and 50° north and south. The farthest north that termites are known to have reached is Hamburg, Germany, where the termite genus *Reticulitermes* was found in a number of warehouses. This major pest genus has also spread north into the southern portion of Canada (Toronto and many other parts of Ontario, and Winnipeg) (Grace, 1989,1990). Some of the genera with older origins still remain in extreme conditions, e.g. *Zootermopsis* in British Columbia, *Porotermes* at the tip of Chile, Tasmania and south-east Australia, and also *Archotermopsis*, which lives at 3000 m in the Himalayas. Different species of the genus *Mastotermes* (Mastotermitidae) were once widely distributed, even in southern England and Europe, in the Oligocene period but are now only found in Australia and Papua New Guinea as a single genus, *Mastotermes darwiniensis*.

Of the Rhinotermitidae, *Heterotermes* is found in the hotter regions of south-western USA, Central America, the Caribbean and India. *Reticulitermes* is a pest in temperate zones, including the USA, Canada, the Middle East, Japan and China, with *Coptotermes* common in tropical America, Africa, the Far East, Australia, the Pacific Islands and Japan. The wood-feeding families, which include the more primitive lower termites, except for the Termitidae, are more common in the Palearctic, Nearctic and Australia south of the tropic of Capricorn. The higher termites (family Termitidae) occur mainly in tropical latitudes with the fungus-growers and humus soil-feeders found in Ethiopian and oriental regions.

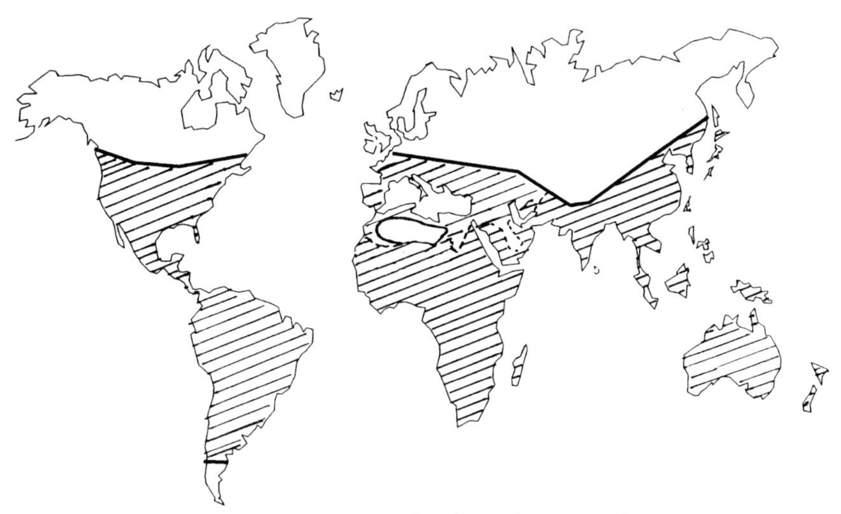

Fig. 17. World termite distribution cross-hatching shows termite presence.

The Macrotermitinae are mainly found in Africa, the Middle East and Asia, and include several pest genera. They are absent from Australia and the Americas, their role being partially taken over by the Nasutitermitinae and the Termitinae, especially in Central and South America where fungus-growing attine ants are also found. It is thought that the fungus-growing termites have never reached Australia but could easily survive there today if introduced. Soil-feeders are also absent in Australia.

PEST DISTRIBUTION

Termites can cause damage to crops, buildings, pasture and forestry as well as to non-cellulose materials such as dam linings and electrical cables. Surface soil mounds can also interfere with ploughing and grazing. Around 150 species of termite have been recorded attacking buildings but only about half cause major damage. Figure 18 shows the genera in which many of the major pest species of termite are found while Figs 19–23 illustrate the distribution of some of these.

The Rhinotermitidae and Macrotermitinae contain most of the pest species. The Rhinotermitidae (especially *Reticulitermes* and *Coptotermes*) are major pests in America, Europe and Asia. The Macrotermitinae (especially *Macrotermes, Odontotermes* and *Microtermes*) are major pests in Africa and Asia (Fig. 19). One member of the Rhinotermitidae, *Reticulitermes* (Fig. 22) is spreading westwards. In 1953, fifteen departments in France were affected by termites and by 1992 this had increased to fifty (Serment and Pruvost, 1991). This same genus is spreading north from the

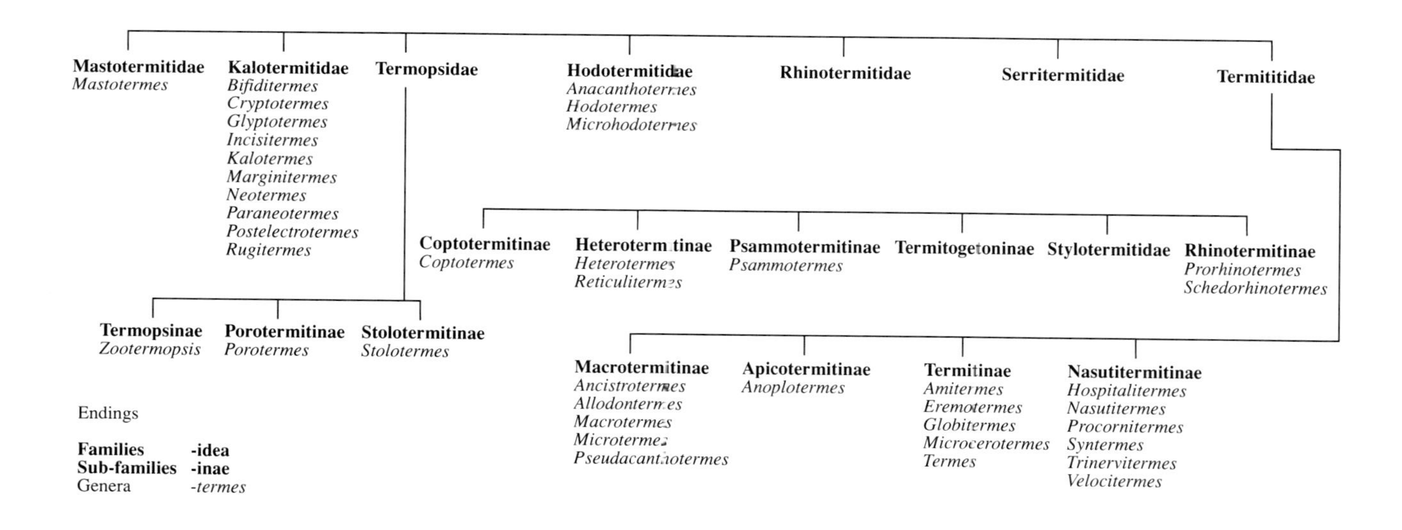

Fig. 18. Families and sub-families of termites with a list of the genera that contain important pest species (where no genera are given, there are no important pest species).

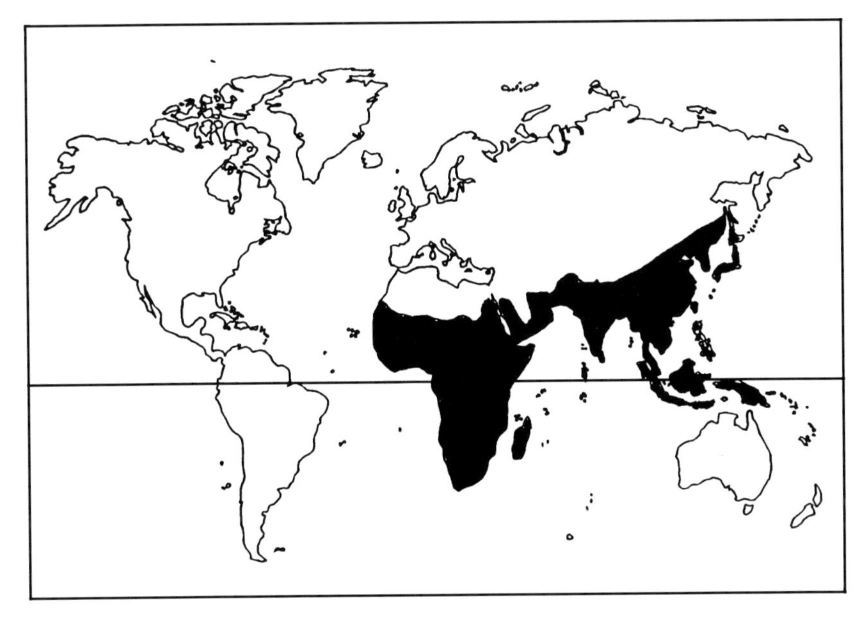

Fig. 19. Distribution of termites in the sub-family Macrotermitinae.

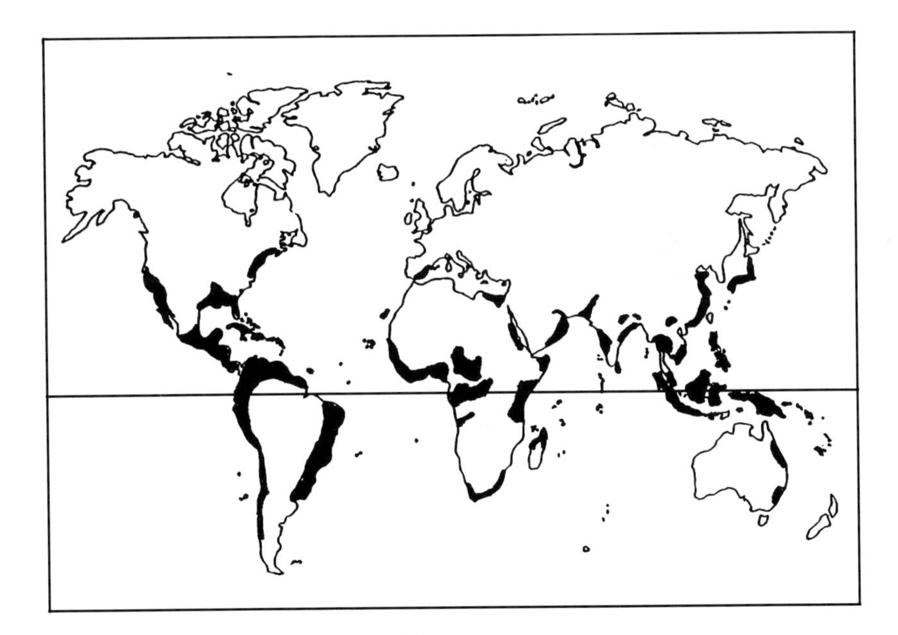

Fig. 20. Distribution of drywood building pest termite *Cryptotermes* spp.

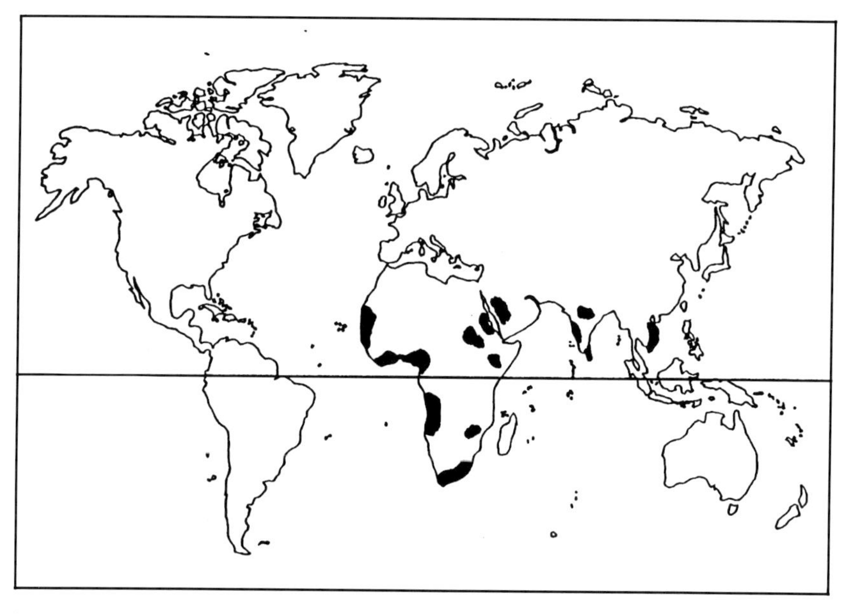

Fig. 21. Distribution of the pasture feeding termite *Trinervitermes* spp.

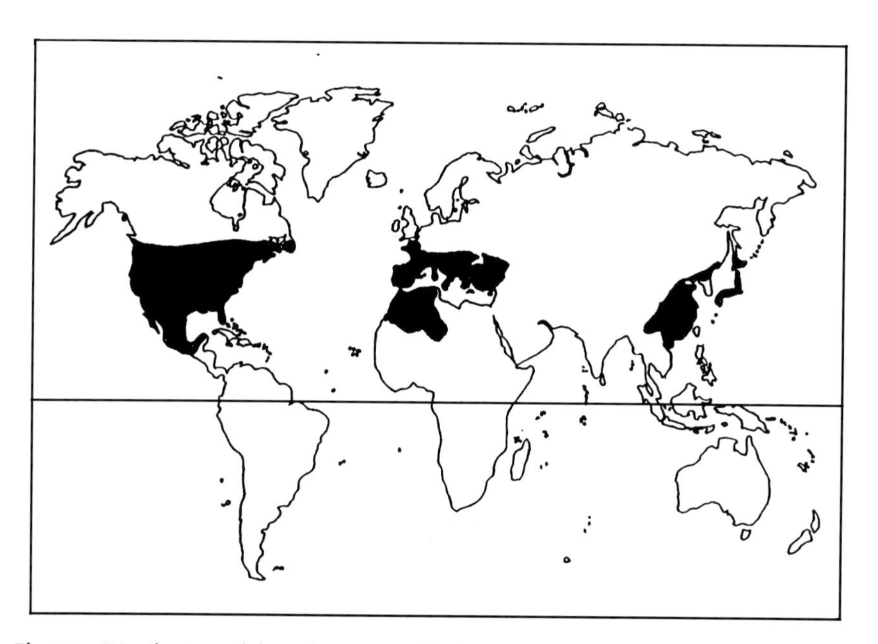

Fig. 22. Distribution of the subterranean building pest *Reticulitermes* spp.

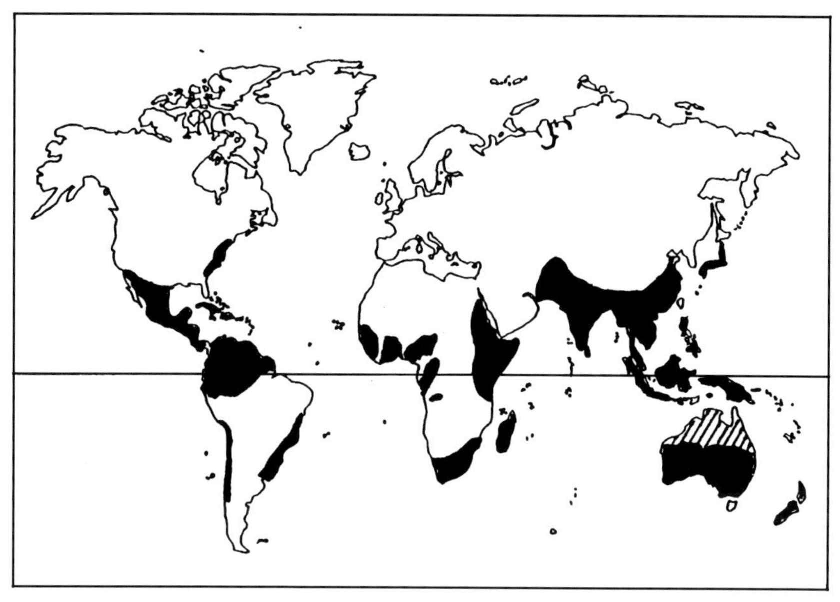

Fig. 23. Distribution of the building pest *Coptotermes* spp. (dark area). The cross-hatched area shows the presence of a major termite pest species, *Mastotermes darwiniensis,* in Australia.

USA into southern Canada (Grace *et al.*, 1989). *Coptotermes havilandi,* an ornamental species, has recently been found in Miami, Florida, USA (Su *et al.*, 1977) where it may have entered by boat at least 10 years ago.

FACTORS AFFECTING DISTRIBUTION

Land Bridges

Drifting of landmasses over time can explain the present distribution of termites but land bridges may also be responsible. The Bering bridge may have helped termites, e.g. *Microcerotermes* from Africa, move through Indomalaya to the New World (Emerson, 1955). Cowie (1989) also suggests that the Arabian peninsular formed a land bridge between Oriental and Afrotropical regions during the Miocene to Pliocene periods allowing movement of termites westwards.

Islands

Some termites, such as the genus *Cryptotermes,* rely on high moisture content within the wood for a water supply and are therefore restricted to living near the coast, rivers or lakes (Fig. 20) (Williams, 1976). Islands would have to be close (a few kilometres apart) for termites to transfer successfully from one to the other. The main method of transfer is inside floating logs (rafting). Success of termite survival in floating logs will depend on the size of the log (amount of food source) and how much has been attacked by the termites. Also the type of wood may affect survival, e.g. soft wood may be more permeable to water than hard and more easily broken. The ideal situation is to have newly developed colonies from alates which have slow colony development times. Although, where supplementaries are also produced, a few workers can still form new colonies. Examples of termites which have been transported in this way include members of the Kalotermitidae (e.g. *Cryptotermes* spp.) and in the Rhinotermitidae, *Coptotermes* is known to have spread in this way (Williams, 1977).

Natural Barriers

Some termites can occupy areas or environments where other termites cannot survive. This may be linked to many factors, including temperature extremes, the kind of vegetation present and the amount of rainfall. Prolonged exposure to low temperature can restrict movement and eventually cause death due to starvation. For termites to survive in cold regions they must compensate for loss of population in winter by increasing numbers during the warm season. Termites are not found at altitudes much above 3000 m. The number of species (diversity) is highest in tropical rainforests, while outside the tropics there are fewer.

Away from tropical regions termite numbers become fewer and the termites less destructive as the climate becomes less favourable, e.g. drop in temperature or increase in aridity. However those termites that managed to adapt and survive in the past still remain in colder regions today. Examples include *Archotermopsis* in the Himalayas and *Zootermopsis* in the American Rockies.

Rainfall can also affect distribution. The percentage of humus-feeders increases with increasing rainfall. In total desert no termites can survive due to the dry conditions and lack of food. The Sahara, for example, acts as a barrier between Ethiopian and Palearctic faunas. However, around oases, irrigated areas, edges of deserts, run-off from hills and semi-arid regions certain termites may survive. Distribution of termites is limited by high temperatures, although the sand termite, *Psammotermes,* can survive in and around semi-arid regions of north Africa and India and other similar districts. *Anacanthotermes* can also be considered a true desert species.

Spread by Humans

Various ways exist by which termites can be transported. Locally made boats, wooden parts of ships, timber, railway sleepers, ornamental plants or nursery stock, and even furniture can all provide a means of carrying live termites. Plant quarantine regulations may not cover termites or inspections may only be concerned with insects on the plant not in the soil in which the plants are kept. Drywood termites such as *Cryptotermes* can survive in furniture anywhere in the world as long as the building has some form of heating.

In parts of America the development of waterways has helped the spread of termites in timber being transported. Many new infestations were introduced during wartime, again through transport of timber as packing cases and for other purposes. *Cryptotermes* was introduced in this way. *Cryptotermes dudleyi* and *C. domesticus* reached India via imported timber.

Survival of termites in Europe depends on mild winters or centrally heated houses. The genus *Reticulitermes* is thought to have arrived in the region south of the river Loire in France in the 13th Century at Charente Maritime before infesting the Mediterranean basin and Paris, where it is a major pest today (Clement *et al.*, 1996). The species *Reticulitermes flavipes* found today in Hamburg, north-west Germany, probably arrived in the late 1930s in pine timbers from the USA. A recent record of *Reticulitermes* surviving in a centrally heated house in Cornwall in the UK is another example of transfer by humans. *R. flavipes* was transported to Winnipeg, Canada, in 1987 where it survived. Current concern in Ontario, Canada, is over increased introduction of termites from Malaysia and the USA in packaging and crates (Myles, 1995).

Chapter 3

Termite Biology and Behaviour

As termites spend the majority of their lives underground, inside timber or nests, the study of their behaviour is not easy. This is especially the case for those termites that are difficult to maintain in laboratory cultures. To gain a better understanding of behavioural responses it is important to identify the kinds of sense organs and chemical secretions possessed by termites. Observations on colonies in the field and in the laboratory give us a better understanding of the characteristics of termite behaviour, which enables us to appreciate the relationships of termites within the environment, and to design control methods where termites are significant pests.

COMMUNICATION

A great deal of termite behaviour is instinctive. Responses are repeated and carried out in sequence unless disturbed by surrounding influences. Unlike many other insects, termites live in the dark, so sensory and chemical communication (touch and taste) are very important.

The benefits of community life are many. Mutual grooming and feeding ensures that all castes are cared for, and that contact and chemical communication is widespread throughout the colony. Different roles exist for different ages or castes within a colony, especially in the higher termites. The termite body, as in other insects, is covered with hair-like sensilla which vary in function, number and distribution according to the termites sensory requirements. The number of sensory sensilla increases towards the end of an appendage involved with contact or detection. Their number and arrangement may also be linked to species and caste. Richard (1969) illustrates the different arrangements of sensilla.

Many sequences of behaviour are stereotyped, i.e. interruption of one part of the sequence will force the termite to begin the sequence all over again. This can be seen in burrowing and building. Each part of the behaviour is triggered by a stimulus that is the result of the behaviour carried out.

Sense Organs

Chemoreceptors

Wood-dwelling termites are exposed to various chemical constituents as they tunnel. New alates leaving the parent nest also have to decide whether a particular type of wood on which they land is suitable in which to make their nest, but may have little choice as their main aim is to reach a place of safety. The effect of the chemical constituents of wood is one reason why some woods are resistant to termites.

Soil-dwelling termites that leave the nest to forage outside or above ground are likely to be exposed to a wider range of chemical stimuli. Termites can distinguish between odours using small peg-like sensilla (basiconic) that have numerous holes on their walls allowing air to enter via diffusion. These are found on antennae, palps and other parts of the mouthparts and are used in the detection of such chemicals as sex and trail pheromones.

Other sensilla used for taste are chemoreceptors which are found on antennae, mouthparts, labrum and other parts of the body (Fig. 24). In some termites the number of chemoreceptors can differ for the different castes,

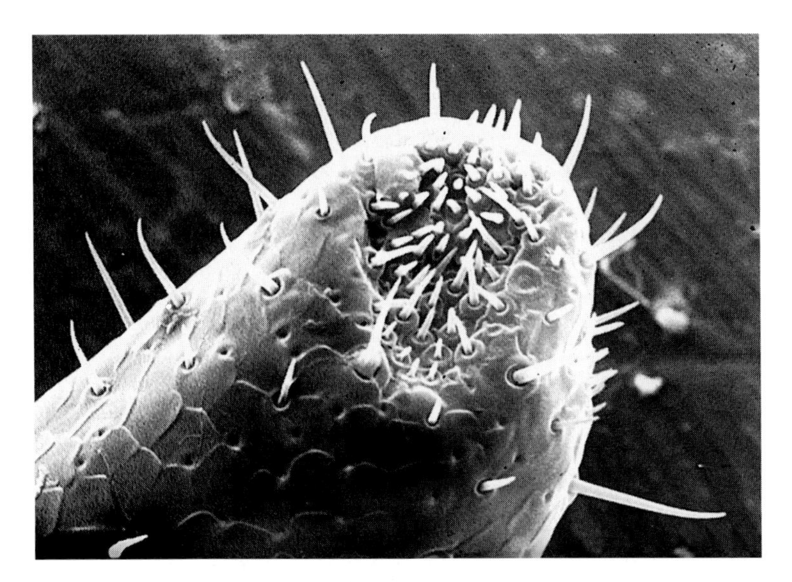

Fig. 24. Tip of the maxillary palp of *Cryptotermes* showing the concentration of different sensilla for taste and touch. Also present is a horseshoe shaped campaniform sensilla.

most being present in reproductives. *Kalotermes* female alates have more chemoreceptors than males. Termites' sense organs are able to distinguish between salt and sugar, as well as between many plant chemicals. Cellulose is an arrestant, causing termites to stop and feed, while glucose and the amino acids proline and lysine are feeding stimulants, causing termites to feed longer and take in more food.

Grooming involves the stimulation of chemoreceptors and mechanoreceptors. Observations show that all parts of termites are groomed, the antennae and legs being groomed from the femur to tarsus. Grooming also appears to be an initial contact behaviour, before soliciting for food.

Humidity receptors are also present on the antennae, and trails and runways can be directed towards a water source. Before colony foundation in some termites, male and female alates often chew off the last few segments of each other's antennae, removing sense organs which may have once been needed while above ground.

Mechanoreceptors

When displaced, sensory hairs on the body and appendages of the termite can register contact through nerve stimulation. Campaniform sensilla, mechanoreceptors which rely on distortion of the cuticle for response, are also found but more often in areas of the body where movement and contact is more common, e.g. at joints. Groups of receptors may be arranged in different ways to detect stress in different directions. Some receptors may be found near the tip of appendages that come into contact with objects. This allows fine monitoring for carrying delicate objects such as nymphs and eggs.

Mechanoreceptors may help to detect material stress. It is generally thought that termites will not attack and weaken timber under stress. Feeding is avoided on wood that is under strain from heavy weights or that is made up of portions held apart like a clothes peg. Termites will also eat cork in bottles but not so completely as to allow the liquid to escape. When termites are inside wood or a plant they leave a thin outer layer and do not tunnel out into the open. This is believed to be linked to air movement through the thin areas rather than the result of termites detecting light penetrating the thin layer. *Zootermopsis* is especially sensitive to moving air and will seal up even the smallest opening.

Termites, especially wood-dwelling ones, are attracted to narrow spaces. This has been seen by using wooden baits with different size gaps between them. Alates also select various sizes of holes to enter after flight. Before alates can enter or dig a hole they may have to fly for a certain period until they lose their wings.

Chordotonal sensilla

Two groups of sensilla (chordotonal) are located in the second antennal segment of termites (Howse, 1970; Kirchner *et al.*, 1994). One is termed the 'Johnston organ' and responds to movement while the others may respond to gravity, linked to the weight of the sensillum. Both work by affecting membranes, which stimulate cilia of nerve endings. Experiments placing termites on a vertical surface have shown that *Microcerotermes beesoni* tends to move upwards and *Neotermes bosei* move down, responses which may be related to their normal foraging movements.

Subgenual organs

Subgenual organs which respond to termites tapping on the ground are found below the knee of the termite, and these have cilia sensitive to vibrations but which only distinguish large differences in vibration. Alarm can be caused by a sudden noise, the presence of an intruder, or the release of an alarm pheromone by a member of the colony. When disturbed, soldiers bang their heads on the substrate. This tapping alerts the colony. In *Zootermopsis* the pulse repetition rate is about 20 Hz (Howse, 1970). Workers also vibrate, but with forward and backward movements. Violent tapping inside wood is not needed as wood acts as a good transmitter. Some alates possess a tympanum but this is believed to be a non-functional evolutionary remnant.

Compound and simple eyes

Termite reproductives (alates) have a pair of compound eyes and a pair of simpler light detectors called ocelli positioned above the eyes on the top of the head. In brachypterous neotenic reproductives compound eyes may not be well developed. Some soldiers may also possess rudimentary eyes.

Vision in termite alates is as good as in many moths and beetles. Alates are photopositive (attracted to light) on leaving the nest, then on landing after forming a pair they become photonegative. Sunlight has been shown to be more attractive than ultraviolet, which in turn is more attractive than incandescent light for *Cryptotermes* species. Termites are sensitive to light intensity. A light intensity of above *c.* 100 candela can cause mortality in *Coptotermes*. More termites are caught in light traps during the full moon rather than during the new moon. Termites are unlikely to detect different colours, but *Bifiditermes* has been shown to have preference for light blue, then pink, greenish blue and then green.

Once the alate starts a life inside wood or soil the eyes start to degenerate. Eyes of workers and soldiers in *Anacanthotermes* and other termites that forage outside the nest in the day do not degenerate as much as those of the imagos. Workers which appear to have no eyes can, however, detect light. One example which supports this is *Zootermopsis*, which seals cavities where light is entering.

Other cuticular structures

The cuticle of termites is often covered with pores through which glands secrete hydrocarbons and waxes that may be used for chemical identification and defence against desiccation and microorganisms. Termites carry their own odour as well as a general nest odour on their bodies and so they can recognize an intruder. Larvae from different colonies of the same species appear not to have a strong odour and can therefore be accepted. Exocrine glands which appear as pore plates sunk into the cuticle have been found on legs of some Rhinotermitidae. Their function is not clear but it may be associated with the grooming of the antennae and legs or for production of a defence secretion against attack by ants on the legs (Bacchus, 1979). Soldiers of some termites possess setae, or bristles, which may have certain functions. In some genera of the Rhinotermitidae these are concentrated at the end of their labrum and can act like paint brushes (see section on defence, pp. 55–60). *Spatulitermes* soldiers from the southwest African and Zambian desert areas have spade-like bristles over their heads (Fig. 25) and *Coendutermes* soldiers from Brazil have many hollow glandular bristles on their heads. The function of these is not clear but possibly may be linked to defence. Alates often have sticky pullvilli (pads) on their feet to allow them to climb vertical surfaces, but this is not always common in all species of the same genus.

Communication Using Pheromones

Pheromones are chemicals which are often volatile and can initiate a response or change in behaviour in other individuals. In a termite colony they are important in communication. A great deal of termite behaviour is stereotyped. Release of a particular kind of behaviour may be triggered by pheromones at different concentrations. Foraging, defence, nest building and alarm trails all involve pheromones. The morphology of exocrine glands that produce pheromones is covered in detail by Quennedey (1975).

Hydrocarbons are present in the wax layer on the cuticle and can act as clues for species or caste recognition. Tergal glands are present in termites. They are not found in alates of Termopsidae, Hodotermitidae and some Rhinotermitidae. *Mastotermes* and *Macrotermes* have glands on tergites 3–10. *Kalotermes* has pairs of glands on tergites 9 and 10. In *Microtermes obesi* the trail pheromone is unsaturated diterpene hydrocarbon and workers and soldiers respond to it. Soldiers of most termites appear more sensitive to trails than workers, and soldiers may recruit more soldiers.

Termites can detect and follow a gradient of trail pheromone to orientate towards discovered food sources (Grace *et al.*, 1988). The trail pheromone (*cis, cis trans*-3,6 8-dodecatrien-1-ol) has been found in the sternal gland of several species of *Reticulitermes* and *Coptotermes* (Ampion and Quennedey, 1981). A similar compound is found in the female sex pheromone of *Pseudacanthotermes spiniger* in the sternal gland. Sternal glands have ductless cells and lack reservoirs in the

Fig. 25. The dorsal view of the front half of the head of *Spatulitermes* showing the arrangement of spade-like sensilla on the nose.

Kalotermitidae, Termitidae and Hodotermitidae. This means the pheromone is transported from the gland cells out of the cuticle. In the Rhinotermitidae, however, secretions from gland cells are stored in a central lumen and released through ventral cuticular ducts (Percy and Weatherston, 1974). At a low concentration this acts as a trail pheromone. Attractant sex pheromones are fanned into the air by wings of alates. The female alate of *Allodontermes giffardi,* in the Ivory Coast, does not fly but climbs to the top of grasses to attract a male. *N*-tetradecyl proprionate in the sternal gland of European *Reticulitermes flavipes* alates attracts and elicits grooming behaviour.

Communication through Grooming

Grooming is one of the most important and common forms of communication within the colony. Grooming with mandibles and palps allows sensory and chemical contact between individuals, as well as the removal of soil, bacteria, fungi and parasites. The queen in the colony is constantly groomed. In lower termites, during food shortages excessive grooming can lead to cannibalism of antennae and legs. The sequence of grooming can be regular. In *Microtermes* the grooming begins at the head and antennae. A considerable time is spent on palps, which may or may not lead to soliciting of food. Legs and antennae are groomed from the proximal to distal ends.

Termites may turn on one side to allow grooming of the lower parts of their bodies. Multiple grooming involving several individuals at the same time is also seen. Termites may present legs for grooming but these may be rejected.

Broodcare

Eggs, when laid by the queen, are coated in an attractive liquid which the waiting workers drink. Eggs may be carried by the workers in their mouthparts (palps and jaws) to other chambers or to a separate part of the royal chamber. They are often held together in a pile by salivary secretions, but may be moved around for cleaning. If the egg pile is large, and the eggs laid first are in the centre, workers will enter the egg pile to remove hatching eggs. Some termites such as *Anacanthotermes* have hatching spines on their head and thorax that enable them to break out of the egg. Larvae, on hatching, remain in the brood chamber with the reproductives who look after and clean them. When the first workers develop and have begun to forage for food, these take on the role of looking after the young. The larvae are translucent with large setae. They can be assisted in hatching by workers who pull off and eat the egg shells. The larvae are then cleaned thoroughly to remove any remains. In fungus-growing termites larvae may be carried and placed on to the fungus-combs. In the event of a serious disturbance from outside or within the nest, the eggs and the larva are carried to another region away from the source.

Effects of Magnetism on Behaviour

Wedge-shaped mounds of some species of Australian *Amitermes,* e.g. *Amitermes meridionalis,* are built with the long axis of the wedge aligned in a north–south direction. (Duelli and Duelli-Klein, 1978). If bar magnets were placed at the base of these mounds in a different direction then the colonies failed. Some *Odontotermes* queens in India position themselves along or perpendicular to the earth's magnetic field. Other effects of magnetism on the direction of foraging tunnels and egg laying have been shown by Becker (1976).

FEEDING

Trophallaxis

Trophallaxis involves exchange of secretions/liquid food between individuals. Termites can pass partially digested semi-liquid food from the crop or secretions by mouth (stomodeal feeding) (Fig. 26), or receive secretions

Fig. 26. A soldier of *Macrotermes* receives food from the mouth of a worker.

from the anus of another termite (proctodeal feeding). This is linked with grooming; that is, the termites are often groomed first which then leads to acceptance of a request for food. Several workers can be involved with the transfer of food at the same time. This is especially important when there is a shortage of food and moisture and is important for the transmission of chemical messages to other termites, throughout the colony and finally back to the queen. The queen is a better donor than the king. The degree of trophallaxis within a colony depends on its size and age, and seasonal variation of food supply. Salivary glands of the termites may vary in size. In *Macrotermes bellicosus* the largest salivary gland size is found in the nursery workers for feeding the young. In the lower termites that have protozoa in their gut, salivary glands are lost just before moult but are regained later by trophallaxis.

When termites feed, other termites are ready to take over their duties. Termites in feeding groups or individual castes can show a form of agitation by moving their bodies quickly backwards and forward repeatedly. If this disturbance is sufficient, a feeding worker will move away to let the intruder feed at the site. Larvae can attract nurse workers by the same method.

Wood Feeding

Termites feed in groups, often in rows or circles. There is a definite organized pattern of feeding which ensures that work is carried out at the same place and with the maximum use of all workers available. In wood-dwelling termites rows of similar sized chambers can be found as well as even tiers of feeding which allow for maximum use of the wood where numbers are great. An aggregation pheromone has been found in the labial gland of the termite *Schedorhinotermes* that leaves a chemical message for other workers at a food site. The presence of a royal pair has been shown to increase feeding intensity.

When eating wood, termites hold the wood fibres between their mandibles and tear pieces off by moving their heads to the side. Where pieces are small these may be carried back to the nest. Major workers, where present, can deal with harder food than minor workers.

Microtermes workers have been seen to carry out sweeping movements with the antennae across the ground in an area where all the wood present has been eaten. Whether this is a form of searching behaviour or a means of gathering up minute fragments of wood on the antennae which can later be groomed off is not known. As the food supply in termite colonies runs out, then progressive moults lead to the formation of alates so that, although the parent colony dies out, other related colonies can be formed in a new food source. This change in the colony to alate production is linked to more encounters as termites search for food, and to changes in feeding rates, which affect the hormonal balance in the colony.

Other Food Sources

Other food sources can include animal fodder, old sacks and palm matting. Even old discarded shoes can be eaten. Close to urban areas the only food source may be domestic animal dung or rubbish. Some termites, e.g. *Psammotermes hybostoma*, may rely on the fibre left in animal dung when no other food is available. Other termites, e.g. *Odontotermes* species, can be considered the 'rubbish tip' termites as they are often found feeding on all kinds of discarded waste. Dung from grazing animals and wild game also can be an important food source for termites (Ferrar and Watson, 1970).

Fungus-comb Consumption

Fungus-growing termites of the family Macrotermitinae eat wood and litter which then pass through the gut and are deposited as pellets in a comb-like structure in the nest (Wood and Thomas, 1989). In *Microtermes* the depositing of pellets is carried out as a specific routine. The termite first examines the area where the pellet is to be deposited with its palps and antennae. It then turns with its abdomen positioned over the spot, turns around again to examine the area and again turns before releasing a pellet. The abdomen is raised up suddenly after deposition, so that no remains of

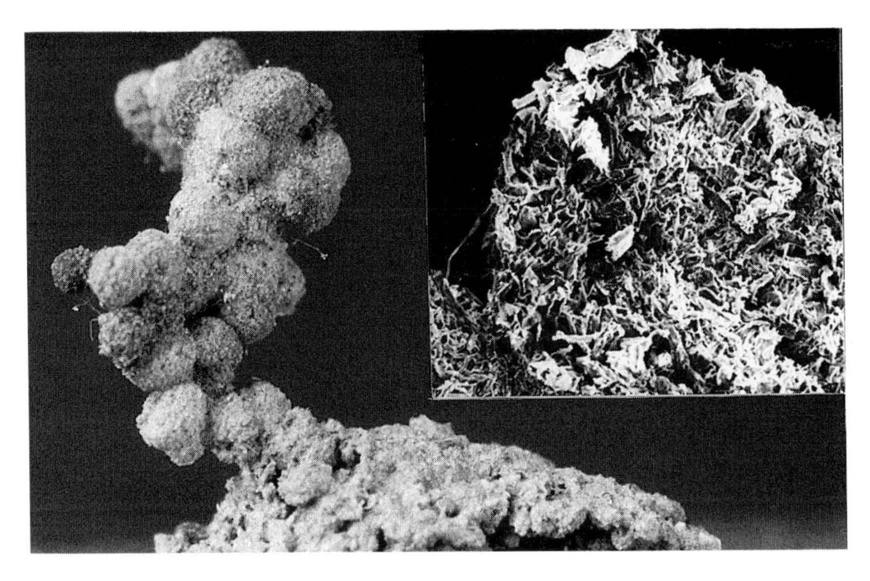

Fig. 27. Initial stages of fungus-comb formation by *Microtermes*. The smaller photo shows a section of one of these faecal balls consisting of chewed wood fragments and fibres.

the pellet are left on the termite. The pellet is released as a soft ball which quickly hardens (Fig. 27). Pellets are arranged initially as pillars connected together to form arches, which in turn are joined together to form an elaborate network of channels and tunnels within a comb-like structure. Within this structure *Termitomyces* fungi, specific to the species of termite present, then break the comb down to simpler materials which can be reingested by termites. Some species of termites, such as *Microtermes* and *Odontotermes,* may choose a site for comb building close to, or within, a food source. This is advantageous in reducing foraging time and therefore the possibility of attack by predators.

Asexual spores of the fungus form an important food store, as do the fungal hyphae in the comb. Major and minor workers, and especially larvae, feed on comb nodules and older workers feed on the comb. *Odontotermes* fungus-comb has been shown to contain the sugars glucose, galactose, mannose, arabinose and xylose, which are thought to serve as binding materials (Agarwal, 1978). Specific areas of fungus-comb appear attractive to termites. The reason for this may be that termites are taking in water containing dissolved sugars from cellulose breakdown of the comb. This is especially useful for soldiers that rely on liquid food provided by workers. The termites do not actually bite into the comb but their mouthparts are in contact with the comb surface. In the case of *Microtermes* this feeding behaviour may last for up to an hour. The comb may offer a place where termites can rest and feed so that they are ready and recharged with energy for further foraging. Termites spend most of their larval life on, or close to, the comb which brings them into contact with food, moisture and other workers. The size of combs may increase at the expense of other combs when food is low or absent, i.e. some recycling of combs can occur. When feeding on fungus-comb, pieces are bitten off and manipulated with the palps and moulded with saliva before ingesting.

Enzymes and Digestion

Food passes to the crop where it is ground, and then passes to the proventriculus where secretions are added from the midgut, and then the liquid is absorbed in the hind gut.

The salivary glands of termites can contain several enzymes, such as amylase and chitinase. Cellulase in the gut can originate from microorganisms present, such as protozoa in the lower termites. A termite cellulase is also produced in the midgut and in minor portions of the foregut.

For fungus-growing termites, e.g. *Macrotermes,* the highest cellulase activity is found in young workers, which eat the cellulase-containing nodules on the comb, so digestion is in the gut. The young workers remain in the nest to process plant material, build combs, eat nodules and then pass food to larvae in the form of a pre-digested liquid. Soldiers are fed older fungus-comb, and therefore only have gut cellulase for digesting cellulose. Old major workers of *Macrotermes subhyalinus* eat old fungus-comb so one would not expect much comb-derived cellulase activity in their gut (Veivers *et al.*, 1991). This, therefore, suggests that fungus-comb is cultivated as a source of nutrients. In addition to the above-mentioned enzymes, a chitinase for breaking down fungal walls and eating nestmates has been found in the Macrotermitinae originating from eating fungus-comb. Fungal breakdown of wood before it is eaten by termites can help to make wood more digestible and provide added nutrients from the fungus itself.

The Importance of Protozoa in Digestion

In the lower termites, flagellate protozoa are important for the breakdown of cellulose in food. Gut protozoa take up wood particles by phagocytosis in the hindgut (paunch) (Fig. 28) where most of the food absorption takes place (Varma *et al.*, 1994). Protozoa can digest hemicelluloses and to some extent cellulose. These are fermented anaerobically by protozoa in the gut to produce acetate and carbon dioxide, hydrogen and methane, which are released. The acetate is absorbed and used by the termites as their oxidizable energy source, and protozoa by breaking down cellulose provide the energy for this. Without the presence of protozoa these termites would die.

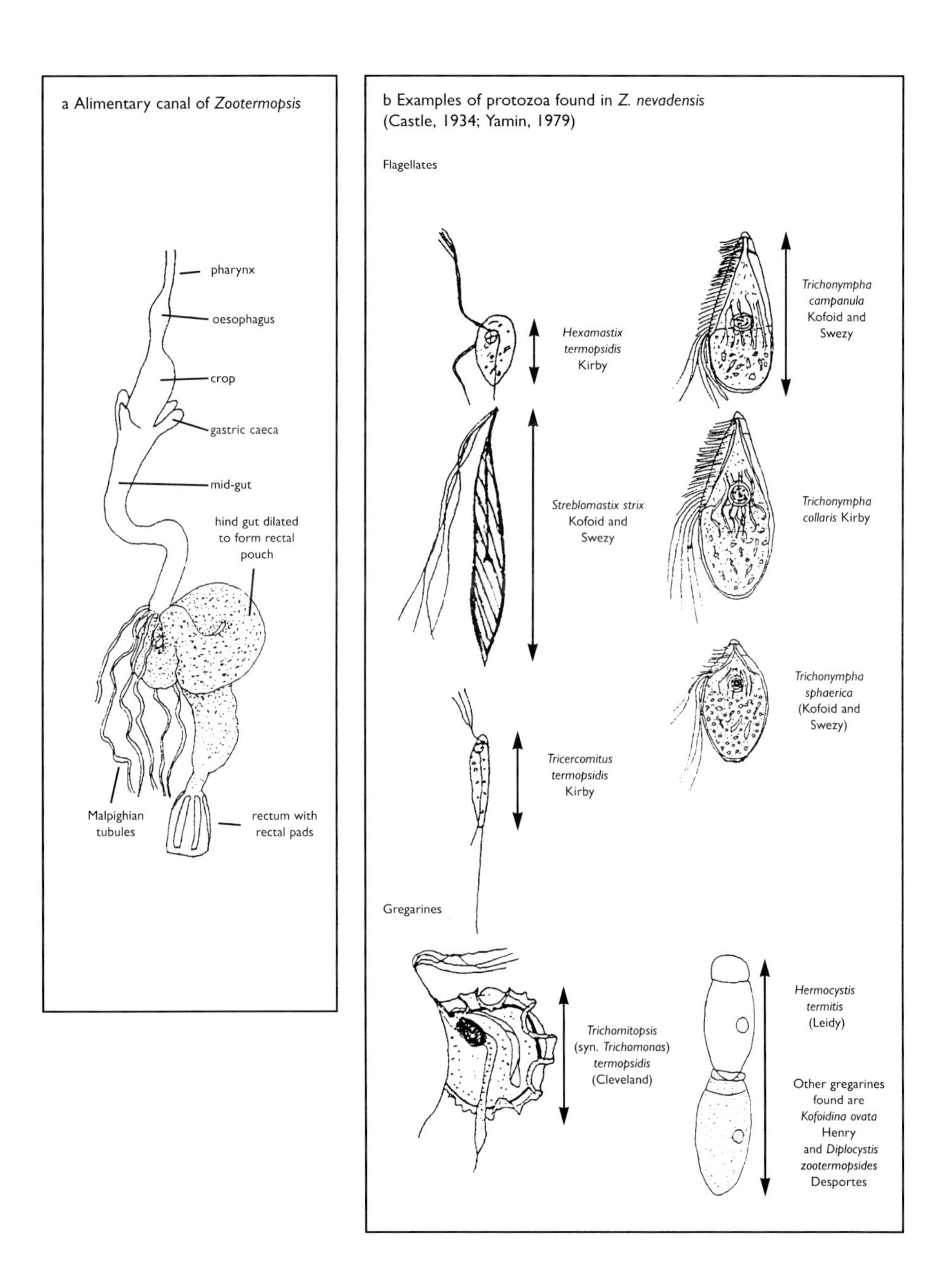

Fig. 28. The alimentary canal of the termite *Zootermopsis* (Termopsidae) with three examples of protozoa commonly found inside the paunch.

Bacteria present in the gut also can help to breakdown compounds to acetate.

Microbial symbionts may also produce a chitinase. Lignin digestion may be initiated by microorganisms anaerobically and also occurs during continual proctodeal transfer of faecal material.

Protozoa may be specific to a particular termite genus. Some protozoa also have bacteria associated with them. Pseudomonad rod-shaped bacteria are found on the *Streblomatrix* protozoa in the hindgut of *Zootermopsis* (Fig. 28). These bacteria act as sensory chemotactic symbionts. *Mastotermes* also has protozoa with associated bacteria. Gregarine protozoa can also be found in the genus *Zootermopsis* (Fig. 28) but these are most likely parasites. Higher termites may also possess protozoa but they are not very important for cellulose breakdown. In the higher termite *Odontotermes formosanus,* from Japan, gregarine protozoa have been found attached to midgut epithelial microvilli and bacterial rods are attached to spines of the paunch and colon.

Nitrogen Sources

Nitrogen is important for the formation of amino acids and proteins needed for growth and survival of termites. A termite diet of plant material contains little nitrogen. Survival may actually depend on access to nitrogen-rich cambium, and it has been suggested that *Zootermopsis* actively defends nests rich in cambium nitrogen.

Nitrogen can also be obtained from fungi present in wood, which is especially common in damp and wetwood feeders. Fungi are also a source in the fungus-combs built by Macrotermitinae. Mycotetes, the small vegetative fruiting bodies of the fungus *Termitomyces* which contain asexual spores and are present on the fungus-comb surface, also provide a good source of nitrogen. They are readily consumed by young workers of fungus-growing termites. Animal dung from wildlife is often eaten by termites, but dung from domestic animals is not eaten. This may be because fungi normally present in wildlife dung are absent from the dung of domestic animals and humans.

Another source of nitrogen which enables termites to survive nitrogen-deficient diets is through the fixation of nitrogen gas anaerobically by bacteria present in the gut. Rates of nitrogen fixation and nitrogenase activity can be seasonal and depend on several factors including food quality (Curtis and Waller, 1995). Many termites produce uric acid, which is stored in the fat bodies as white deposits. This can be broken down into ammonia which can be used as a nitrogen source (Breznak, 1983) .

Cannibalism and Entombment

Cannibalism forms another important source of nitrogen, which is normally low in a cellulose-based diet. Nothing is wasted in the termite colony. When the primary queen of the southern African termite *Amitermes hastatus* cannot survive any longer, the workers lick her to death until all that is left is a shrivelled skin. Alates, if they do not leave the nest, are often eaten by the workers. Before flight they often segregate away from other members of the colony to avoid this occurring.

The setal patterns, or number, on some termites may help them to recognize friend or foe. If the termites moult to form replacement reproductives (supplementaries) and lose abdominal setae they may be attacked. Elimination of excess supplementary reproductives involves detection by smell. The presence of haemolymph from an injury causes cannibalism which is triggered by the fatty acids present (Dhanarajan, 1978). The amino acids, aspartic acid and tyrosine have also been shown to elicit attack by *Coptotermes formosanus* (Yaga, 1972). Cannibalism is common in termites where the food source is low, and often appendages are bitten off. In the Kalotermitidae, one may often find wing-pad or leg scars resulting from attack by other workers. This is not found in the Rhinotermitidae. In the Kalotermitidae evidence of regeneration after amputation exists. Cast skins including inner linings of foregut and hindgut at moult and egg shells are eaten by termites. In some termites, even alate partners once in the nest eat off the end of each others antennae, which may have become redundant.

If dead termites are not eaten they may be sealed off or become part of the nest in mounds of soil. Poisoned or dead termites are often repellent after a certain time. In drywood termites, such as *Cryptotermes,* these are either sealed off in small chambers or completely covered in sealing material to isolate them from the colony and prevent the spread of pathogens and mites. In soil-dwelling termites the dead workers may be incorporated into the structure of the building, such as pillars or walls. In the Macrotermitinae definite cemeteries have been found, often near the junction of two colonies, where the colonies have been in conflict over territory.

WATER REQUIREMENTS

Water is essential for termite survival. Movement outside the colony in high daily temperatures for many of these soft-bodied insects could mean rapid desiccation and death. Exceptions are those daily foragers living in humid forests, e.g. *Hospitalitermes,* and members of the family Hodotermitidae in grassland, which are protected by a hardened, pigmented cuticle.

Water is required for body functions, building nests, soil tunnels, regulating temperature and feeding other termites and the young. Moisture can be obtained from many sources. Drywood termites rely on water held inside the wood which depends on contact with damp soil or high air humidity. Water is also obtained from the metabolic breakdown of sugars in food. These termites have elaborate rectal pads for reabsorbing water in the faeces leaving a dry seed-like pellet. Dampwood termites, as their name suggests, are found inside wet wood and they therefore have ready access to water. The mode of feeding of the different species of termite can affect the degree of absorption of water by pieces of wood.

For termites living in semi-arid regions or areas with seasonal water shortages access to the water table is important. Tunnelling depths of 10 m have been recorded for *Microtermes*, 40 m for *Psammotermes* and even more for some of the *Macrotermes* (Lepage *et al.*, 1974). Cavities or chambers built by termites close to the water table may provide water by having wet soil in their walls.

Termites have water sacs (salivary reservoirs) inside their bodies for storing water. In soldiers, these can also be used for defence where chemicals are stored inside. In workers, water present in these sacs is especially useful when moistening soil particles for building. These 'tankers' are especially important in semi-desert areas where surface soil moisture content is very low. In *Odontotermes* nest building activity slows down at 14% moisture, and optimum soil moisture is 22–26%. Organic matter in large amounts in the central part of a nest can help to retain water, also carton (chewed wood plus saliva) in some nests has a high water-holding capacity. In dry areas it is especially useful for termites to have access to the soil surrounding soil particles, as this is where water can be found. The hypopharynx in some termites, e.g. higher termites such as *Macrotermes*, has a line

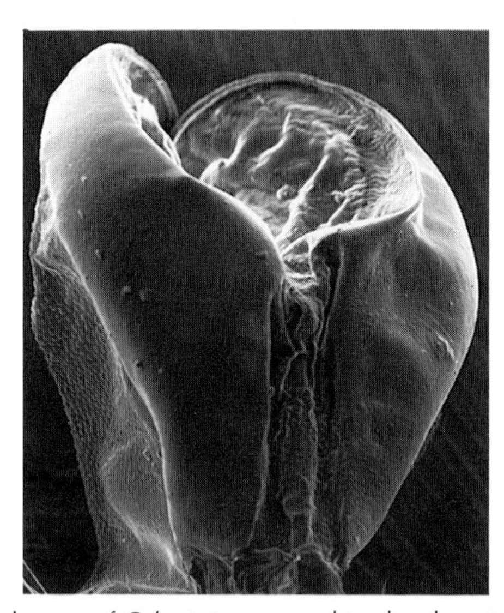

Fig. 29. The hypopharynx of *Odontotermes* used to absorb water from the soil and fungus-combs.

of backward pointing hairs (Fig. 29). These, together with the pumping action produced from changing the shape of the hypopharynx, help to suck water away from the soil (Lys and Leuthold, 1994). This water also supplies minerals to the termite. The water-holding capacity of runways of some termites, e.g. *Odontotermes,* is higher than in mound soil. This can enable foraging for longer periods in drier conditions outside the nest.

Fungus-combs in mounds of the Macrotermitinae also supply water through fungal metabolism, which can be taken in by termites eating the comb. Observations on *Microtermes* also suggest that termites remove water from the surface of the comb, much as they do for water around soil particles (see fungus-comb consumption in previous section). Both water removal from soil particles and especially from the fungus-comb may be the only times when the termites are least active, and a major time for rest and recovery. This rest period could be equivalent to periods of inactivity seen in lower termites living inside wood (Maistrello and Sbrenna, 1994).

DEFENCE

Different mixtures or proportions of constituents of the hydrocarbons found on the cuticles of different termites are detected by sense organs on the termites' antennae. Colony odour which is slowly volatized from the epicuticular waxes may be one of the methods used for recognition of friend or foe in the termite colony. Different species of termite, or termites from different colonies, try to avoid one another by blocking off galleries to prevent conflict. However, where food resources are restricted large colonies can take over the foraging areas of smaller colonies. Also, more aggressive termites such as *Coptotermes* will replace other termites such as *Reticulitermes.*

The defence of a colony against intruders is mainly carried out by the soldier caste (Deligne *et al.,* 1981). In some species the workers are much larger than the soldiers so that individual soldiers are less effective and therefore the workers also have a role in defending the colony. Breaking up a gallery or nest causes soldiers to appear and stand on guard to attack intruders while the workers repair and seal galleries. Soldiers may also be present during foraging. In *Nasutitermes exitiosus* large soldiers are present with the foragers while small soldiers are near the nest. The reverse is also seen in other termites, e.g. *Macrotermes michaelseni.*

In mounds in the Macrotermitinae. the royal cell has solid walls with a line of small exit holes which are easy to defend. Mechanical defence (mandibles) and chemical secretions are found in soldiers of many termites (Prestwich. 1983).

Mandibles

Many different kinds of defence adaptations in the soldier caste have made them unable to feed themselves and dependent on workers within the colony for this. The simplest soldier mandibles are for crushing. These are large, strong and thick. In the Kalotermitidae, Hodotermitidae and Termopsidae they are smooth or have a few teeth. Large mandibles are no good for holding small predators as they escape but the workers in the colony can deal with these. For cutting, weak elongated mandibles are found, as seen in *Basidentitermes*. Some mandibles are long, and when crossed over can be flicked back so as to cause a sideways blow and throw the intruder to the side. Examples of these are seen in *Neocapritermes, Capritermes* and *Termes*. In *Pericapritermes* only the left mandible flicks out so that the predator has to be on the left. Very sharp mandibles are found in some termites for piercing and in addition chemicals are also found.

Head Shape

Some soldiers have plug-like (phragmotic) heads used to block up entrances so that predators, such as ants, cannot enter. One example is in the genus *Cryptotermes* (Fig. 30), where soldiers are known to remain blocking the gallery to intruders, and will die in this position. A similar method is found in *Campanotus* ant species.

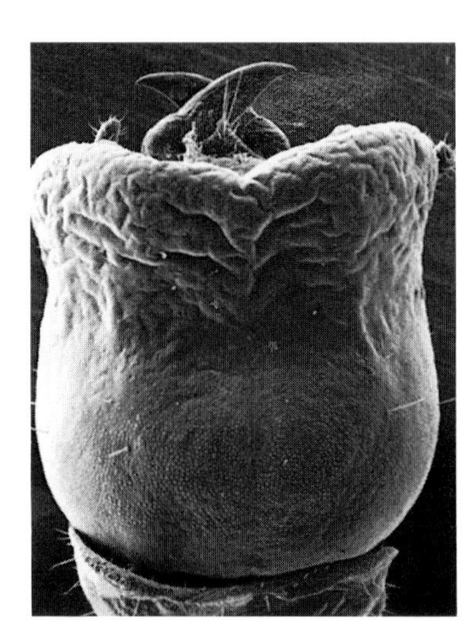

Fig. 30. The plug-like head of *Cryptotermes* used to seal tunnels against entry of predators.

Locking mandibles across a gallery can also stop intruders. The shield-like structures on *Cavitermes* soldiers and frontal structures on soldiers of *Spinitermes* help to block galleries. Some termitophiles (other invertebrate groups that live with termites, see Chapter 4) are present in the nest and also may help to repel intruders.

Chemicals

Secretions from various glands in termite soldiers have a defensive function. The larger the gland, the greater the reliance on chemicals. Cephalic glands, which tend to be the largest, open at the base of the mandibles and produce a white sticky secretion. Frontal glands, where developed, open towards the front end of the termite head. Volumes of secretion in major soldiers, which have larger frontal glands, can be much larger than in minor soldiers, e.g. in *Macrotermes subhyalinus* the gland secretion is five hundred times that of minor soldiers. The best example of a large opening to a frontal gland is in the fontanelle present in *Coptotermes* soldiers. The white glue produced from this species consists of a lipid in a mucopolysaccharide. *Mastotermes* produces benzoquinones which, when mixed with saliva, form tanned protein. *Odontotermes* soldiers produce a white sticky substance from the mouth, while soldiers of *Globitermes sulphureus* produce a yellow liquid, which is also released when the frontal gland, which is in the abdomen, ruptures (Bordereau *et al.*, 1994). This rupturing (dehiscence) also occurs in *Serritermes serrifer* soldiers but the liquid released is from a salivary reservoir.

Some members of the family Rhinotermitidae, such as *Schedorhinotermes,* have bristles located at the end of the labrum which they use as a brush to daub the glandular defence secretions on to the enemy. In *Rhinotermes* soldiers, the clypeus is lengthened with a groove. This allows the glandular secretion to run down to the labrum so that it can be spread by the palps or a labral brush on to an enemy.

The soldiers of the sub-family Nasutitermitinae have long narrow nose-shaped heads with an opening of the frontal gland at the tip (Fig. 31). In *Nasutitermes princeps,* once the soldiers immobilize the enemy then the workers join in. Soldiers can fire several times with the secretion, which hardens in the air. It is believed that they must use auditory or olfactory cues to aim in the right direction, as they do not have eyes. Release occurs only when the enemy approaches. Sticky threads are wiped off the nose on the ground after firing. Eisner *et al.* (1976) gives a detailed account of the defensive behaviour towards predators by the nasute termite *Nasutitermes exitiosus.*

Soldiers of the sub-family Termitinae have various different means of chemical defence including glues, oils, dehiscence and irritants (Table 3; Fig. 32). The secretions of Nasutitermitinae often contain terpenes.

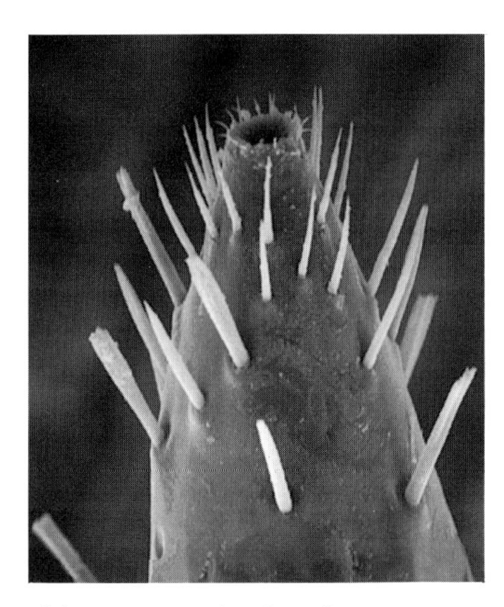

Fig. 31. Tip of nose of the nasute termite, *Spatulitermes.*

Monoterpenes can also act as an alarm pheromone with the diterpenoid used as the carrier. *Lacessitermes* has a mixture of monoterpenes, sequiterpenes and diterpenes. *Trinervitermes geminatus* has monoterpenes with pinenes, camphene and limonene. The cuticular glandular secretions found over the surface of the head of *Nasutitermes* may also have a defensive role.

Alarm pheromones are often volatile resinous terpenoids. A monoterpene α-pinene is a major component in the alarm pheromones of some *Nasutitermes.*

Worker Defence

In *Microtermes,* major workers are more aggressive than soldiers. Another form of defence is autothysis as described by Sands (1982). Thus, if a worker of *Alyscotermes kilimandjaricus* (Apicotermitinae) bites an ant, the termite's abdomen splits open and two droplets of liquid emerge to cover the ant. Other termites that do this include workers of *Astalotermes quietus* and *Ruptitermes.*

Natural Defences

Termites often live in the soil or damp wood where there are numerous fungi, bacteria and other microorganisms. Termites produce antibacterial

Table 3. Examples of defence mechanisms shown by different termite families and sub-families.

Family/sub-families	Soldier defence mechanisms
Mastotermitidae	Mandibles, salivary gland irritant (quinones)
Kalotermitidae	Mandibles, head plugs, tunnels
Hodotermitidae	Mandibles
Rhinotermitidae	Mandibles, frontal gland secretions (toxic and repellent)
Coptotermitinae	Proteinaceous mucopolysaccharide glues from large frontal opening
Heterotermitinae	Terpenes
Prorhinotermitinae	Contact poisons
Psammotermitinae	Chemicals, unknown
Rhinotermitinae	Contact poisons from elongated labum with brush tips
Termitogetoninae	Chemicals, unknown
Termopsidae	Mandibles
Serritermitidae	Mandibles, dehiscent workers
Termitidae	Mandibles and numerous chemicals
Apicotermitinae	Mandibles and dehiscent workers
Macrotermitinae	Mandibles and salivary defence (corrosive oils and quinones)
Nasutitermitinae	Reduced mandibles, elongated heads with pointed tip as gun for terpenoid glues, irritants or oily lactones
Termitinae	Mandibles, terpenoid oils and irritants (workers may also be aggressive)

Source: Adapted from Deligne *et al.* (1981)

secretions from glands in the cuticle. Termite saliva can suppress the growth of fungi and, with the removal of contaminant fungi within the colony, can help to reduce the spread of harmful fungi. Fungi can also be killed by their passage through the gut. The Macrotermitinae use soil moistened with saliva to construct the walls of chambers that contain fungus-combs. When the chamber is opened contamination occurs. Removal of infected/dying termites from the colony or isolation/burying (Fig. 33) (Pearce, 1987) can prevent spread of infection. This is one reason why low doses of biological agents have been unsuccessful as the uninfected individuals seal themselves off from those which have been infected.

Another method of defence seen in the alates of some termites, e.g. *Trinervitermes*, is that they can feign death so that they are not selected by predators. To protect the queen from attack by a predator she can be carried off by the workers to another place of safety. The vast network of tunnels which may be restricted to different areas of a nest can often confuse intruders or allow sealing off of particular regions.

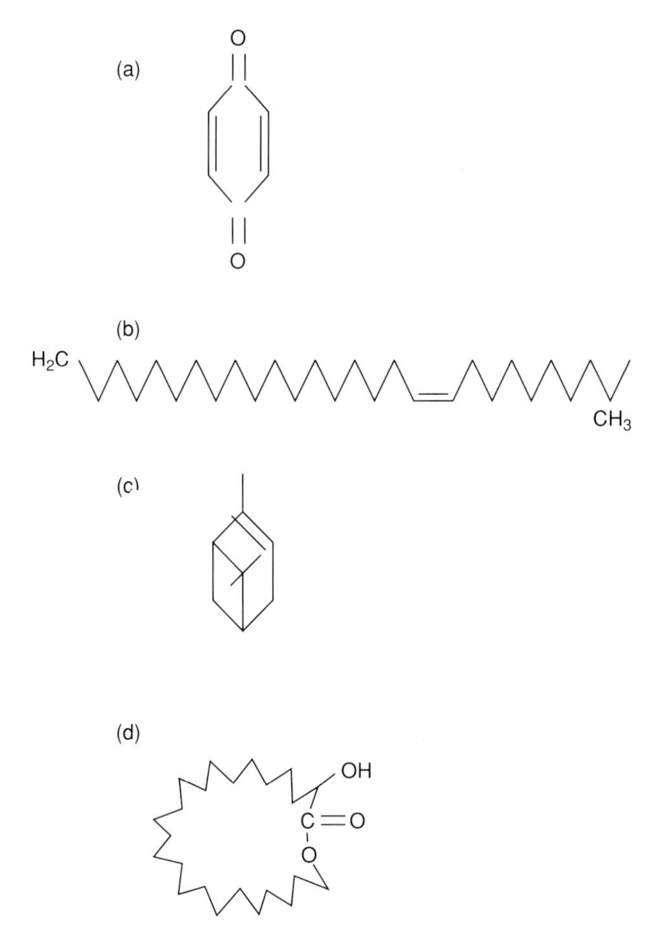

Fig. 32. Examples of the structure of defence secretions of termites: (a) simple quinone (common in Mastotermitidae and Termitidae); (b) alkenes (Termitidae); (c) α-pinene (Nasutitermitinae); (d) a monocyclic lactone (Armitermes).

FORAGING

Galleries

The most efficient foragers are able to build over non-woody material, to forage over long distances and have efficient defensive castes. Tunnels and runways have to be larger than the largest soldier's head.

Soil galleries are made by moistening the soil with saliva, moulding it into a soft pellet with the mouthparts and applying it with a rocking movement. Soil runways, as found in *Macrotermes,* are usually made with a high proportion of clay and some sand particles moistened with saliva. Figure

Fig. 33. A dead work termite of *Cryptotermes* sealed off from the rest of the colony to prevent infection.

48 (Chapter 6) shows how soil infill is important. Soil runways (small tunnels of soil) can be built over the outside of plants. On tall trees these can extend several hundred metres up into the branches. Soil sheeting, large areas of soil under which termites can travel, is also produced by some termites and can completely cover plants, or other woody material (Plate 5). Soil sheeting is often thinner than soil runway material and is, therefore, more fragile. In very sandy regions runways are often unstable and are not built to vertically great heights. *Coptotermes* and *Nasutitermes* use paper like faecal carton for runways.

In some *Macrotermes* species, if a major worker finds food more major workers are recruited. Minor workers may also then be involved in the transfer of food back to the nest.

Anacanthotermes, Schedorhinotermes, Macrotermes, Fulleritermes, Hospitalitermes, Trinervitermes and *Nasutitermes* forage in the open air. *Hospitalitermes sharpi* can have files 300 m long with soldiers on the outside (Jander and Daumer, 1974). Another species of *Hospitalitermes, H. umbrinus,* also leave the nest in a continuous stream. They leave in the evening and return the next morning carrying food balls which can include fungi and bryophytes (Collins, 1979b). Harvester termites, such as *Hodotermes,* can move out in long lines. *Trinervitermes biformis* can leave nests in thousands at sunset, workers in the middle and soldiers at the edge, and return at sunrise.

Foraging can be a long haul – at least 50 m in some termites, e.g. *Odontotermes* or *Macrotermes* spp. – or only in the nest, as in drywood termites. A distance of just over 100 m has been recorded for subterranean foraging by *Coptotermes formosanus* (Su and Scheffrahn, 1988).

Territory will be influenced by the number of galleries and amount of gallery use, as well as the ability of the entire colony to move between logs or soil depending on conditions. Foraging tunnels under soil increase in number and depth if a food shortage occurs. Termites in the field and in urban areas tend to forage randomly and eventually appear to concentrate at a few sites.

Hodotermes forages both in the day and at night. It has eyes and uses light from the sun and moon for orientation. Chemical trails are more important when light intensity decreases. The concentration of the trail can indicate the distance from the nest. *Anacanthotermes* forages most at midnight and early morning in summer and winter, with the reverse occurring in spring. In rainforests, where there is high humidity and low light intensity, termite foraging can continue throughout the day.

Foraging distribution

The number of *Reticulitermes flavipes* in an urban area in Florida has been estimated as 2361 m^{-2}. These can be as many as ten million African *Trinervitermes* individuals per hectare compared with a few million for *Cubitermes* and *Macrotermes*. *Microtermes* in Nigeria have been found at up to 50 cm below the soil surface in cultivated regions but can move deeper into the soil in the dry season (Black and Wood, 1989). From baiting experiments in India, termite activity was found to be greater in maize than in perennial crops, and lower in annual crops, no-tillage areas and grassland.

Trails

Foraging trails are made by the termite's sternal gland, which is pressed on to the ground. This gland is less developed in younger workers. Secretion is made through pores into a cavity formed by a segment in front. The secretion is released from between the segments only when the region containing the glands is pressed against the ground. Campaniform sensilla register the opening size and, therefore, the amount of chemical released. *Mastotermes darwiniensis* has three sternal glands in the middle of the 3rd, 4th and 5th sternites. In *Hodotermes* and *Trinervitermes* they are on the 3rd and 6th sternites. In *Trinervitermes bettonianus* the sternal glands are vestigial in soldiers, very large in reproductives with a lot of trail pheromone, but smaller in workers. In *Stolotermes* and *Porotermes* (Hodotermitidae), they are found on the 4th sternite. In Termopsinae, Kalotermitidae, Rhinotermitidae and Termitidae they are found on the anterior part of the 5th sternite (Noirot, 1995).

A strong trail will cause the termites to produce wider galleries, though trails are less important where soil runways are used as guides. *Reticulitermes santonensis* can perceive space and overcome obstacles with new galleries. The concentration of the trail increases with the number of workers on the trail and therefore affects the locomotion rate and the direction. A normal trail leads to the nest so the food trail needs to be more attractive to encourage other workers to follow. Trails are also used for recruitment for repair of runways or nests. Different castes can have different quantities of trail pheromone. In *Nasutitermes costalis,* trails initially attract soldiers, then as the concentration increases, workers are attracted.

Trail chemicals

For most species of *Reticulitermes* and *Coptotermes* dodecatrienol-1-ol is the primary chemical for the trail. In *R. flavipes* the long-lasting component lasts one year while a highly volatile one lasts for only 15 mins. In *Anacanthotermes ahngerianus* the trail chemical is not effective after 12 h. Individual termites can regulate the amount and composition of the trail. Glycols, which are present in the ink of ballpoint pens, have been shown to attract species of *Heterotermes, Coptotermes, Microcerotermes* and *Reticulitermes.* Some aldehydes and amino acids are formed during wood decomposition by fungi; these attract termites and could also be used in trails.

NEST BUILDING

Termites produce a wide range of nest architecture as will be described in more detail in the next chapter. It has been suggested that nest building initially evolved from a defence reaction where faeces were used to seal off intruders from nests (Fig. 34). This is still evident today in some wood-dwelling termites. *Cryptotermes* alate pairs build a lattice over entry holes once they are inside, which they then fill with faecal pellets. *Zootermopsis* also uses liquid faeces to build up barriers in and around the nest, which is often a response to intruders or air movements. Grassé (1967) put forward the theory of 'stigmergie' whereby a few termites start to build simple structures and at a certain point in construction, another stage begins and other termites are encouraged to join in.

Termites taken straight from the field will often build structures straight away if soil is provided. An area is examined first by the termite, then a soil pellet is placed in that region and moulded into shape with the mouthparts. Pellets are then built up into pillars and then joined at the top to form arches. A certain number of pellets, or a certain height, may trigger off the formation of an arch. Arches are built at a height of 4–5 mm for *Cubitermes* or 5–6 mm for *Macrotermes.* The final orientation and connection of arches may depend on the spatial orientation of these structures detected by the termites. Another possibility is that different concentrations of pheromones produced during building guide the builders.

Pheromones present in the saliva of termites, and also in the fat body of the queen, have been implicated in building. Palmitic acid from the queen's fat body is responsible for the regular arrangement of pillars and arches around the queen in *Macrotermes,* which eventually produces the queen

Fig. 34. Termites sealing tubes with mud to prevent entry of termites from another colony.

cell. For surface builders, nests, as with runways, are built after rain and often at night when it is cooler. Building activity is often irregular – a period of building may occur, followed by a lack of building activity for several months.

In *Odontotermes,* large workers are usually involved in gallery expansion and some in nest expansion. The small workers are important in the repair of the nest. Rapid repair is essential for soil-inhabiting termites to prevent attack by predators, which are constantly patrolling. Soil may be taken from other parts of the nest or runway to achieve this. In *Macrotermes bellicosus,* minor workers are predominant in spontaneous construction and in attendance on the queen. The predominant role of the soldiers is to guard the workers who are carrying out emergency construction.

If a *Cubitermes fungifaber* nest is damaged, the termites can move underground and build a new nest at a distance away. It has been suggested that the plan of the nest is innate, i.e. all individuals know it instinctively. Becker (1974) showed that *Heterotermes* in three adjacent containers would only build walls in the outer two containers. This influence could pass through glass and polystyrene, but not aluminium, plates. He suggested that the termites themselves produce an electrical energy field. Drywood termites produce other structures within their nests. *Cryptotermes dudleyi* builds support pillars inside cavities in wood and uses sealing material for patching over exposed areas (Pearce, 1987).

Chapter 4

Nest Systems

NEST TYPES

Different kinds of nests are shown in Fig. 35. They can be inside wood of trees or buildings, subterranean, above ground as mounds, or arboreal (in trees). Although fairly constant in form for many species, the appearance and kind of nest can vary with environmental conditions, such as moisture availability and locality. In colder regions no mounds are built, the termites remaining underground. In sandy regions they may again be built underground and, in arid areas, may be smaller to reduce water loss.

Some nests may first become established in a tree then move to the ground. *Coptotermes lacteus* nests in *Eucalyptus* but when the colony has outgrown the timber, mounds are produced. *Amitermes evuncifer* in West Africa can be subterranean in dry areas and above ground in wet regions. Nests may have several other subsidiary nests so that the termites can move around according to food source or environment. Different colonies of the same species also can have different nest forms.

Nest Foundation

On landing, the female alates of most termite species raise their abdomens and release a pheromone to attract males. A tandem pair is then formed and the pair look for a crack or gap in wood, or a moist area of soil in which to burrow. A small chamber is produced where the eggs are laid. The first workers produced forage for food and, as the colony grows, nursery chambers are used to house newly hatched larvae.

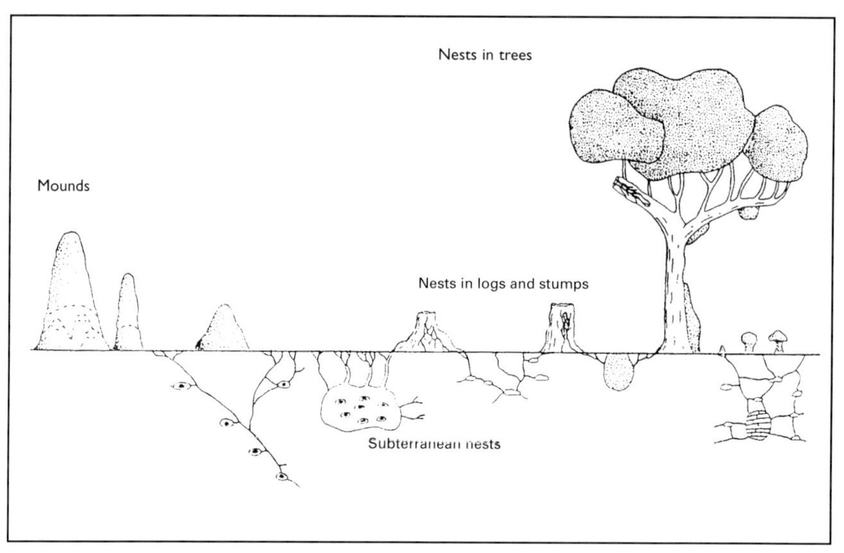

Fig. 35. Termite mounds, nests in wood and subterranean nests.

Wood-dwelling Termites

Small colonies living inside a piece of wood (one-piece nest – Abe, 1987) may never grow beyond a few hundred individuals, but some can reach a few thousand (Plate 6). *Mastotermes,* members of the Rhinotermitidae (e.g. *Coptotermes* and *Reticulitermes*) and the Termitidae (e.g. *Macrotermes*) can have nests containing several million individuals. *Coptotermes* nests have been recorded as having nine million termites. Wood-dwelling termites have fewer competitors than the mound or subterranean termites, but as the wood becomes eaten out with increasing colony size, more predators can enter. As with wood-dwelling cockroaches, galleries are parallel with the grain of the wood. Termites can live entirely within the food source and move to other sources of wood. Drywood termite faeces are dry seed-like pellets that are thrown out of holes made in the wood, which are resealed afterwards. Wood can be totally isolated from the ground in a tree, building or even a boat or in contact with the soil, where the lower half may be damp. In hot climates termites are found mainly in wood close to the ground. Wood can be timber or living, which then provides a water source.

Nests Associated with Trees

In rainforests, where runoff of rain could destroy nests, some termites have developed rain-shedding structures (Plate 7). *Constrictotermes* produces rain-shedding ridges on the tree to direct rain away from the nest.

Schedorhinotermes lamanianus has many subsidiary nests in neighbouring trees connected by galleries. Nests of *Odontotermes* spp. may be found inside trees, e.g. date palms in Sudan, while *Mastotermes* nests have been found inside hollow trees in Australia.

Arboreal nests often consist of undigested wood mixed with wood fragments and saliva. In tropical America and the Far East carton nests of the genus *Nasutitermes* can be found in trees or poles (Plate 8). In these regions the ratio of carton to soil can vary for different nests.

Subterranean Termites

These pose a greater pest threat than drywood termites. In urban areas, nests of *Coptotermes, Reticulitermes, Mastotermes* and *Schedorhinotermes* commonly extend under buildings. These nests can be compact or can be a set of diffuse chambers. *Hodotermes* nests, which may have several large chambers (hives) of thin papery structure arranged as parallel cells, can be found as deep as 14 m below the soil surface. *Coptotermes* nests are made of carton and commonly found inside dead wood/trees or in warmer regions may be found inside mounds with a thick outer layer of soil and carton (Plate 9).

Coptotermes may have several satellite nests so that the colony can move nearer a food source or to a more favourable environment. In young colonies of other genera, where the queen has not yet greatly increased in size, queens are also able to move with the colonies.

Some nests may be subterranean or on the surface in a similar locality, e.g. *Odontotermes feae* is found underground in India but in mounds in Bangladesh. Soldierless termites have small nests found in diffuse galleries or in nests of other species. Fungus-growing termites such as *Microtermes* may have very diffuse nests with a network containing small chambers (Plate 10).

Mounds

These are usually constructed after the colony has developed below the ground. *Amitermes, Cubitermes* and *Procubitermes* nests are often mushroom-shaped with drip tips to help remove rain quickly. In dry areas the nests of *Cubitermes* species often do not have this umbrella cap. Magnetic mounds of *Amitermes* in Australia are wedge-shaped and reach 5.5 m in height with a width of only 0.9–1.2 m. Some *Macrotermes* mounds, e.g. *Macrotermes goliath,* can reach 12 m across, at least 4.8 m high and can number as many as 17 ha^{-1} (Figs 36, 37). In *Macrotermes* in Nigeria, two kinds of mounds may be made by similar species, in similar conditions (Collins, 1979a). One has a spiral plate supported by a central pillar, the other does not have this arrangement. The spiral plate has downward directing vanes that are covered in salt deposits (Plate 11). Darlington (1985a) has described the structure of mature mounds of *Macrotermes michaelseni. Odontotermes* mounds are larger in India than in Africa. Species differences also occur between mounds in India: *Odontotermes obesus* mounds are conical while *O. wallonensis* are dome-shaped.

To resist the washing away of soil by rainfall, a large proportion of clay is needed to build large or tall mounds. Inside a mound there is a usually a cell that houses the royal pair. This is often harder and thicker than the rest of the nest. In *Macrotermes* this cell is extremely hard and has perforations to allow the entry of workers and soldiers.

For termites with a definite royal chamber, a relationship exists between queen length and the size of the royal chamber. In many termites a relationship also exists between nest size and the population of termites inside. Macrotermitinae are not found in Central America and Australia. In these areas where other genera have mounds, they do not have a central clay royal cell but large mounds with a honeycomb system, often constructed of carton-like material.

Microcerotermes can have nests inside wood or can build mounds with hard carton walls, often mixed with soil as with *Coptotermes*. Dead mounds can be recolonized by the same species or by several other species. Nests of some Macrotermitinae in the tropics are built as small mounds under the floors of houses, where the termites can cause serious damage.

Ventilation

Mound-dwelling termites need air to survive. Figure 38 shows the internal network of ventilation channels for a *Macrotermes* mound. Mounds can have several different kinds of ventilation systems. *Macrotermes* mound ventilation has been described by Darlington (1987). Three kinds of ventilation have been identified. In the closed type of mound, termites rely on convection within the nest through small pores through the walls. In the many-holed nest type (Bissel) air is drawn in at the base which cools the mound, while in the single-hole type (Marigat) air is drawn out from the top and in from openings in the surrounding soil. Diffusion currents drag air in at the sides and up the central chimney, the effects of wind have also been implicated for mounds with a large chimney, where air is drawn out at the chimney mouth.

Variation in mounds can also be seen for *Odontotermes* species in India. *Odontotermes obesus* and *redemanii* mounds have no surface openings while *O. wallonenesis* has open chimneys. Chimneys in *Odontotermes* mounds may also be important in ventilation, but this is not believed to play a major role in temperature regulation. Figure 39 shows the small ventilation chimneys of an *Odontotermes* species from Zimbabwe, which protrude just above the ground.

Fig. 36. Tall, cathedral-like mound of *Macrotermes bellicosus* from Ghana, Africa.

It is thought that termites are able to regulate air flow through the nest by opening and sealing the air shafts that affect the temperature within the nest. As nest volume increases, the number of shafts also increases. On the outside of some *Apicotermes* nests there are elaborate ventilation systems seen as simple pores or slits, which connect to circular galleries within the nests (Fig. 40).

Fig. 37. Chimney-like mound of *Macrotermes subhyalinus* (now *Macrotermes jeanneli*), Kenya.

Food Storage

Harvester termites, such as *Hodotermes,* store food in chambers so that they need not forage every day (Plate 12). *Anacanthotermes* stores hay, seeds, thorns, leaves, wood and bark in small chambers along the galleries.

Several of the Macrotermitinae that have fungus-combs have chambers or areas of the nest for storing collected material. Chewed wood and other vegetable matter is collected by some *Macrotermes* and *Odontotermes* species. Plant material stored near the top of the nest allows gases to diffuse to the surface. Where food gets wet, it may be brought to the surface to dry by some termites, e.g. *Anocanthotermes. Globitermes* in South-East Asia stores balls of vegetation within the nest.

Fig. 38. Skeleton of cement representing the runways and ventilation shafts present inside a *Macrotermes* mound.

The inner walls of some termite nests, e.g. *Microcerotermes* and *Nasutitermes,* can act as a food source. Balls of chewed plant material are kept with the fungus-combs are found in *Pseudacanthotermes* nests, but these are used for prefermentation.

Fungus-combs

The Macrotermitinae are fungus growers. Fungus-combs may be in separate chambers or stored in a large chamber sometimes bigger than a football, weighing several kilograms (Plate 13).

Fig. 39. Small ventilation funnel produced above the nest of *Odontotermes* from Zimbabwe.

Workers build fungus-combs by depositing faecal pellets of chewed, partially digested wood (mylospheres). Figure 27 shows the initial construction of a comb and a section through a single faecal ball showing fine pieces of chewed wood. Sands (1960) describes the initial fungus-comb construction for *Ancistrotermes*. A species of fungus belonging to the genus *Termitomyces* is cultivated on the comb by the termites. This breaks down cellulose and lignin in the comb to simpler compounds, which are then reingested by the termites. Fresh comb is continually deposited on the top, while the older comb is eaten from underneath. As well as providing a food source for worker termites, combs are essential for the young of the colony, which are often placed on the comb after hatching. These feed on the mycotetes (fruiting bodies) of the fungus.

Combs can act as a reserve food store for termites, especially at times when food is scarce, e.g. in cultivated areas between the time of harvest and new planting when little food is available if crop residues are absent. Agarwal (1978) and Darlington (1994) give details of the nutritional benefits of fungus-combs.

Newly formed colonies need to produce new fungus-combs. Some alates carry spores from the comb in their gut before leaving the nest. Others have to rely on their first generation of new workers finding spores from the mushrooms that have grown on the soil from fungus-combs beneath the surface (Johnson *et al.*, 1981b). The timing of the production of these mushrooms on the soil surface is closely linked to when the termites

Fig. 40. A section through the nest of *Apicotermes lamanianus* showing the elaborate arrangement of horizontal chambers.

fly, thus ensuring that spores are available for the new workers. Successful colonization by new alates is therefore dependent on finding an area where nests are present, so that mushrooms and spores will be found by the new workers. After rain some termites, e.g. *Odontotermes* in Africa, will bring fungus-comb to the surface, which then produces mushrooms.

The fungus-comb in the nest contains many other species of fungi, but only *Termitomyces* spp. is encouraged to grow by the termites. If the comb is removed from the nest, other fungal species, e.g. *Xylaria* spp., rapidly grow on the comb.

Some fungus-combs are built close to a food source ensuring the maximum use of available food. This is seen in some *Microtermes* species where combs can be found just below the soil surface close to a food source. Fungus-combs of *Odontotermes* are also found inside the trunk of date palms in Sudan.

Sphaerotermes comb contains nitrogen bacteria suggesting that workers need bacterial action for comb breakdown before eating it.

TERMITOPHILES

Many nests offer shelter as well as food for termitophiles (the termite itself may be a food source for a predator). Where this relationship exists for the benefit of the termitophile, then it can be considered a commensal (eating the termites' food store); a saprophage (eating the termites' waste) or a parasite (taking food during termite trophyllaxis). Where there is a symbiotic relationship between the termitophile and the termite present, or the organism is harmless, then the organism can be considered an inquiline. Termites can also often gain from having termitophiles in the nest, especially where they help to defend the nest from other predators.

Invertebrates

Termitophiles, which have a close relationship with termites, are more commonly found in tropical rainforests than in drier regions. The presence of termitophiles in a nest indicates the complex social organization of the termite colony.

The most common inquilines belong to the beetle family Staphylinidae and the dipteran family Phoridae. Staphylinidae are often found in Nasutitermitinae nests while Phoridae are found in both Nasututermitinae and Macrotermitinae nests (Disney, 1994). Larvae of the tineid moth *Passalactis* are common in nests of *Schedorhinotermes* and are also found in their runways.

Termites themselves can also be inquilines in other termite species nests and non-aggressive ants exist successfully in termite nests. Cohabitation with another species of termite often happens as the initial colony dies out or when the soldier caste is low in number. Where parts of the nest are no longer used, new termites can colonize but remain separate from each other.

Ahamitermes, Amitermes, Incolitermes and *Termes* live in other termite nests and can feed on carton. *Cubitermes* nests appear very suitable for secondary occupation (Dejean and Ruelle, 1995; Okwakol, 1991). Mounds also are very suitable places for alates to land and burrow to form new colonies. Inside *Odontotermes* nests in Pakistan one can find

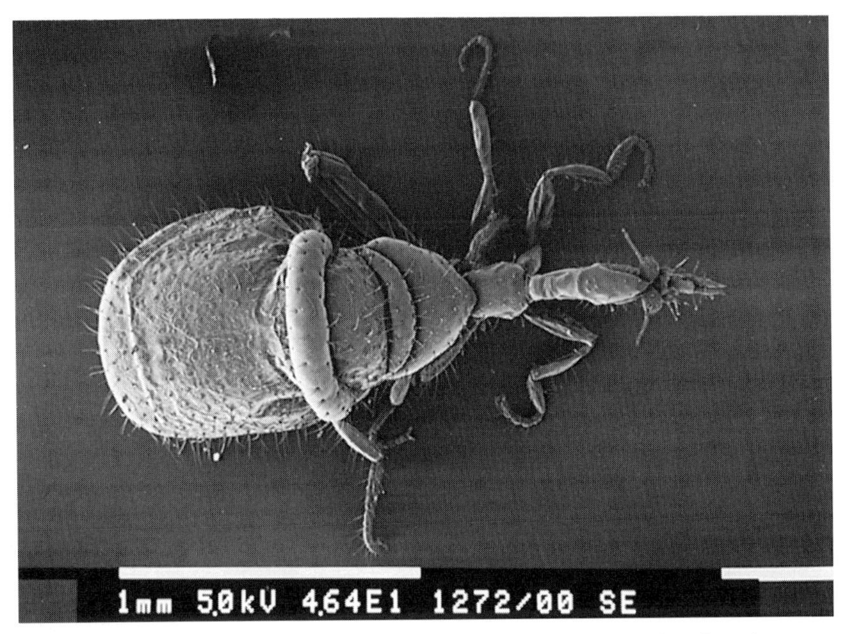

Fig. 41. The female termitophilous inquiline *Clitelloxenia assmuthi* (Phoridae, Termitotoxeniinae).

Hypotermes, Speculitermes, Macrotermes and *Microtermes* living in separate galleries.

To become a successful inquiline a species must adapt methods of being accepted by the host. These methods include having the host colony odour on their bodies and ensuring they have no wings so that they cannot easily be picked up and carried out of the nest. Disney and Kistner (1995) give a good description of one of the more interesting groups, the Afrotropical Termitoxeniinae (Fig. 41). *Termitoptocinus* (Staphylinidae) larvae actually ride on the back of the host (*Nasutitermes*) which ensures that they keep close contact and have access at feeding times. Some inquilines appear like termites having enlarged abdomens. Attractive secretions are often produced and many inquilines have special appendages, or possibly secretory pores (Fig. 42), for this purpose. Some inquilines depend mainly on termite secretions for survival and can even solicit food from the queen, e.g. Staphylinidae with *Grallatotermes* queens. They may also eat termites. In *Drepanotermes* nests, commensal Tenebrionid beetles may not interfere with the termites but feed on stored food such as dried grass and seeds. Bees can make a nest within active termite nests.

Nests of some of the Macrotermitinae can provide breeding sites for phlebotomine sandflies. These carry the flagellate responsible for leishmaniasis, so it is important to locate the breeding sites and destroy them.

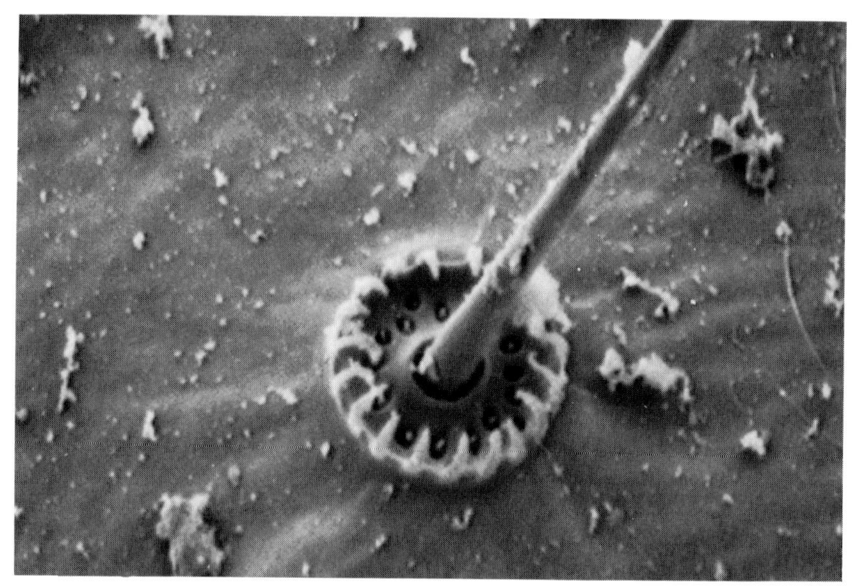

Fig. 42. A solutrichome bristle with termite attractive secretory pores on the body surface of *Necperissa* spp. Tingle (Phoridae, Termitotoxeniinae).

Vertebrates

Some vertebrate inquilines, such as lizards and snakes, are predators of termites. Mounds provide shelter for many animals that feed on termites as part of their diet. Constant temperature, humidity and shade provided by the mounds help in incubation of the inquiline's eggs. The South African monitor lizard lays its eggs inside mounds and the Brazilian crocodile places its eggs on top or on the side of termite mounds. The golden shouldered parakeet (*Psephotus chrysopterygius*) from Australia, also known as the 'antbed parrot', excavates a hole in wet termite mound soil during the wet season. It then lays its eggs in a chamber. The mound, as well as incubating the eggs, also protects the parrot's nest from danger. In Africa, the red faced lovebird (*Agapornis pullaria*) also nests in mounds, and pygmy parrots (*Micropsitta* sp.) nest in arboreal termite nests. In Gabon, kingfishers and woodpeckers also excavate nests in live mounds. In South America the brown throated parrot (*Aratinga pertinax*) can make deep tunnels into mounds to nest (Alderton, 1992).

The lesser dwarf shrew, *Suncus* sp. in Africa, has been found to inhabit dead *Trinervitermes* mounds. In many countries large mounds can also act as rubbing posts and territory markers for many animals.

Chapter 5

Termite Ecology

The presence of termites in a region can depend on various factors, such as soil and vegetation type. Climatic features and water availability play an important part in termite survival. Daily and seasonal changes in these factors also affect termite distribution.

SOIL TYPE

The activity of termites can disturb the soil profile, affect the soil texture and redistribute organic matter (Lee and Wood, 1971; Wood and Sands, 1977). This is more important than any changes in the chemical properties of soil caused by termites. As there are few earthworms in African savannah, termites are important for recycling matter.

Living in soil and mounds or moving under soil runways provides protection for termites in that it isolates them from temperature extremes and moisture loss. Runways and mounds are more stable if clays are present in the soil. The brood chamber in a nest may have the largest clay content. Soil particles inside a *Macrotermes* nest can be a quarter of the size of those on the outer part of the nest. More *Macrotermes* mounds are found in well-drained areas, while very few are found in poorly drained areas.

Sandy soils have very low organic matter content and fewer species are present, especially of the humus-feeding termites. Desert species of Hodotermitidae have deeper nests than in other regions.

True desert-dwelling termites such as some *Amitermes* and *Psammotermes* species prefer sandy and gravelly soil if given a choice. Other termites, such as *Trinervitermes,* are absent in extremely dry arid conditions but prefer semi-arid and wet, open areas. In desert regions, subterranean termites enhance water infiltration and return top soil. The presence of vegetation, or litter, reverses this by decreasing infiltration. Some good references on methods for studying the ecology of termites include Nutting and Jones (1990), Jones (1988) and Grace (1992).

VEGETATION TYPES

By modifying the soil profile, termites can cause changes in vegetation. Certain termites can adapt to a wide range of soil and vegetation types. For example, the area around a *Macrotermes* mound, owing to the constant turnover of soil by termites, may remain clear of vegetation. However, certain plants that are termite resistant due to chemical content or high levels of silica, grow on mounds.

A large termite colony, e.g. of *Macrotermes,* in a savannah habitat can remove more than one tonne of vegetation per year. When a colony dies out or only a small part of a mound is occupied, grass covers it and eventually shrubs develop (Plate 14). In savannah, *Macrotermes,* by moving soil from below ground, creates different layers of soil. Where there is a rich, well-drained grassland, e.g. in the Ivory Coast, humivorous and fungus-growing termites are common. In areas of poor drainage and vegetation these are absent and grass feeders such as *Trinervitermes* and some *Macrotermes* spp. are common. Often, farmers grow crops on mounds. This is advantageous in areas that are flooded, such as paddy fields in Thailand.

Underground nests of *Anoplotermes pacificus* are perforated by the roots of plants, which are eaten by the termites. This has been shown to be beneficial for stimulating new growth.

Primary and Secondary Forests

Regenerating woodland is highly susceptible to termite colonization. Termites can initiate reforestation and affect the kind of vegetation that eventually grows.

In forests, especially during the rainy season, plant debris is digested and incorporated into the soil. Soil feeders are abundant in humid forests of Africa and South America. In wet areas of Africa the importance of the fungus-growing Macrotermitinae declines. As areas of low, open vegetation give way to dense forest, the species richness increases as there are more wood-feeders. Secondary forest also has high species richness due to the large amount of timber present from the previous cover. However, by opening up gaps one loses groups such as the soil feeders *Coptotermes* and *Nasutitermes,* and an increase in savannah species occurs.

In the lowland forest in Sarawak, high numbers of soil-feeding termites are found, while in the continuous forest of Amazonia, a greater species richness is found than in scattered forest. The diversity of termites in Amazonian forests has been studied by Martius (1994). In Nigeria, removal of trees and bushes followed by cultivation resulted in loss of *Macrotermes, Cubitermes, Trinervitermes* and shallow nests of *Coptotermes.* Some subterranean species such as *Microtermes* increased in numbers, as they could move down into the soil to avoid the effects of surface disturbance (Black and Wood, 1989). Clearing primary rainforest in Malaysia and Indonesia for cultivation allows termites to survive where tree stumps remain. One example is *Coptotermes,* which leaves the remains of an old stump or fallen tree when the food supply becomes low to destroy newly constructed buildings. The effect of disturbance on species richness in Cameroon has been studied by Eggleton *et al.* (1995).

Desert

In desert regions, vegetation would remain on the soil surface if it were not broken down by termites. *Heterotermes aureus* in the Arizona desert can remove wood at a rate of 78.9 kg ha^{-1} $year^{-1}$. The greatest attack by termites on dead wood occurs after rainfall. Infiltration then increases, which in turn supports the vegetation present. *Hodotermes* in arid regions of South Africa is important for nutrient recycling. The growth of crops in semi-desert regions can mean the plants may be under water-stress at certain times. Termites in these areas need to have access to water, so the plants become more vulnerable.

BENEFITS TO THE ENVIRONMENT

The problem with termite activity, especially that of mound builders, is that it may create nutrient imbalances, i.e. nutrient-rich (organic and mineral) and nutrient-poor areas. Farmers will then grow crops in the nutrient-rich land, where the growth of crops is better, but this is also where termites are abundant.

In areas where both termites and earthworms are present a balance can exist. In some areas termites are dominant in dry seasons while earthworms are more active in wet seasons.

Nitrogen

In *Macrotermes* mounds, nitrogen and phosphorus cations build up and carbon is lost as carbon dioxide and methane. The royal cell in *Trinervitermes* nests can have greater levels of nitrogen than the surroundings. The soil dumps of *Hodotermes* can have five times as much nitrogen as surrounding soil. Termites can also fix nitrogen themselves (see Chapter 3).

Minerals

Mineral accumulation caused by termite activity is important, especially in desert regions where one may find calcareous masses or soluble salts accumulating inside nests. Asian *Anacanthotermes* accumulate salt in their tunnels and mounds.

The queen cell consists of very hard soil and can accumulate minerals such as silica, aluminium, calcium, iron and magnesium. Chromium, vanadium and precious metals from underground deposits may be brought to the surface by the termites with soil, as a result of mound building. In the wet season, water is forced into the *Macrotermes* mound. It then evaporates leaving the mound rich in elements. Ferric oxide, aluminium oxide and especially calcium carbonate are found in *Odontotermes* mounds.

The calcium present in mounds comes from plants and is probably used as an adhesive for soil particles. Chambers containing fungus-combs can also help with concentrating salts. Russian *Anacanthotermes ahngerianus* nests are rich in water-soluble salts. Phosphorus and potassium salts have a greater accumulation in the top of the nests. In the Okavango Delta in Africa, calcite crystals are left by evaporation of water and brought to the surface when mound building occurs.

The accumulation of minerals in areas of Africa in which soil mineral content is poor can provide a salt-lick for elephants, which eat the termite mound soil. Some materials can remain locked in mounds and are only released when the mound is destroyed. For *Heterotermes,* nutrients are left in workings for months or years, but with *Gnathamitermes* they are released in weeks or months by heavy rains. When fire occurs, mounds are less affected than surrounding areas and grass grows first on these, followed then by shrubs.

Fungus-combs provide protein, inorganic carbon, minerals and small amounts of calcium and magnesium. Soil runways or galleries contain minerals that may be richer than mound material, having high concentrations of calcium, magnesium and potassium.

Water Infiltration

Termite tunnelling can affect overall water movement in soil. Galleries increase the amount of air and water in soil, and improve soil texture by mixing with subsoil. Fine fractions in mound soil have better water-holding capacity than surrounding soils. By increasing the porosity of the soil, the termites also allow entry of other soil organisms and plant roots.

In Africa, water holes may arise from the weathering of mounds and animal tunnelling and rubbing so as to produce clay pans. The presence of termites can also affect the microtopography of soils, i.e. the formation of high and low landscapes. This will affect overall infiltration rates as well as the fertility of the soils.

Methane Production

Methane production is an important contributing factor to the 'greenhouse effect' causing global warming. Methane gas is emitted during the decomposition of organic material by termites. Methanogenic bacteria are important for anaerobically reducing carbon dioxide from wood decomposition to methane. The amount of methane produced is smaller in the Macrotermitinae than the lower termites that have protozoa in their gut. Soil feeders also produce more methane than grass feeders.

Methane-producing bacteria are associated with protozoa or other fauna in the hindgut. Workers produce the most methane and the amount produced increases with size of termite. In *Coptotermes* and *Reticulitermes* workers, bacterial symbionts are found in the paunch (attached by a hold-fast) and also in the midgut between epithelia (Yoshimura and Tsunoda, 1994; Yoshimura *et al.*, 1994). For *Zootermopsis,* methanoic bacteria are possibly symbiotic with *Trichonympha* protozoa in the gut.

It has been suggested that termites could produce at least a fifth of the world's methane. These estimates are extrapolated from estimates from different genera of termites which vary in the amount of methane production. Estimates based on *Nasutitermes* in Amazonia showed that they produce more than termites from other regions. Based on the values found, it is unlikely that termite methane production has much impact on greenhouse gases and therefore global warming (Martius *et al.*, 1993). Most of the wood- and humus-feeding termites, which are responsible for much of the methane production, live in rainforest (Martius *et al.*, 1993).

Gas analysed from Australian *Coptotermes* mounds was also shown to include carbon dioxide, hydrogen, nitrous oxide and chloroform. Levels of hydrocarbons and chloroform were higher than surroundings in winter months with little difference in the summer (French, *et al.*, 1977).

ENVIRONMENTAL FACTORS

Daily and seasonal factors affect termite activity, distribution and population growth. Moisture is the major factor, closely linked to temperature, that affects termite activity. Changes in environmental conditions cause changes in termite behaviour, e.g. avoidance of extremes or increased intensity of normal activities such as nest building in cool, wet conditions.

The spatial structure of colonies depends on environmental conditions. It is important, as it is for other insects, for termites to have access to water. Rainfall, dews, irrigation and the position of the water table provide access to moisture. Over-watering can be detrimental to termite activity. Temperature extremes experienced by termites are influenced by the level of shade provided by various crop types, the ability of the termites to move underground or build soil runways and the presence of shrubs and other objects, e.g. stones and litter.

Some termites are more tolerant to environmental factors. This can depend on the size or degree of sclerotization of the cuticle as well as on adaptations linked to their normal habitat. *Psammotermes* has a higher tolerance than *Heterotermes. Coptotermes* has a greater tolerance than *Odontotermes,* which in turn is greater than *Microcerotermes.*

The presence of a food source is also important, whether it be indigenous shrubs or crops. Some termites (e.g. Macrotermitinae) have reserve food supplies for times when food is short or the climate is too harsh for foraging. Plant damage from pests other than termites, wind, weeding or animal feeding or trampling can encourage termite attack. Some of the factors affecting termite presence, as well as examples of different predators of termites (including man), are given in more detail in the following sections.

Rainfall

Rainfall is the trigger for external building activity and for the release of reproductives from the nest. Alates may not fly if rainfall is low or absent. If there is a lack of rain some termites, e.g. *Macrotermes,* dig down deeper to the water table. Rainfall is accompanied by a rapid change in temperature, humidity and pressure that acts as the trigger for flight, as does the actual sound caused by falling rain drops. Heavy rainfall reduces termite foraging activity.

Humidity

Increased temperature is not the main reason for movement of termites below ground, but a fall in humidity and soil moisture is. With a drop in humidity termites can move to a region of lower temperature. However, the humidity can remain high inside soil runways enabling termites to move into dry areas. If a water table is present close to the surface, termites are also less affected by climatic changes. Kalotermitidae are more resistant to water loss than other families of termite. Survival varies between species, castes of the same species and the same species from different geographical areas (Khan, 1980).

Salt Tolerance

Some termites have a higher tolerance to saline conditions than others. *Coptotermes heimi* is more tolerant than *Odontotermes obesus* and *Microcerotermes championi.* Termites are absent in extreme saline conditions next to the coast. In other areas the build up of salts brought from the subsoil to the surface can prevent plant growth, and therefore termite presence. Where irrigation with fresh water reduces the salt content of the soil and salt-tolerant plants can grow, termites will move in. *Reticulitermes* is known to survive in soil in brackish water regions on the east coast of Virginia.

Temperature

Termites of different genera, or different species of the same genus, can have differing temperature tolerance, for example, *Coptotermes formosanus* has been shown to have a higher temperature tolerance than *Reticulitermes flavipes* (Sponsler and Appel, 1991).

Temperature and moisture affect plant communities, which in turn affect termite presence. In exposed areas in direct sun termites are often below ground or inside nests at midday and early afternoon, when temperature is at its peak. However, they can come to the surface at this time if large objects are present that provide a thermal shadow (Ettershank *et al.*, 1980). In semi-desert regions where little vegetation is present *Psammotermes* is often found underneath small rocks. The thickness, colour and dimensions of the object are important. Large-diameter objects are most attractive as they provide greater shade and moisture. The temperature difference at a few centimetres deep caused by a shadow can be detected by termites. Termites move to the cooler edge of the object as the rest becomes hotter.

The kind of plant cover also affects soil temperature and termite foraging (Fig. 43). Fields of cereal crops offer less protection than other plants and shrubs. Foraging is greater in orchards than in cultivated areas, which

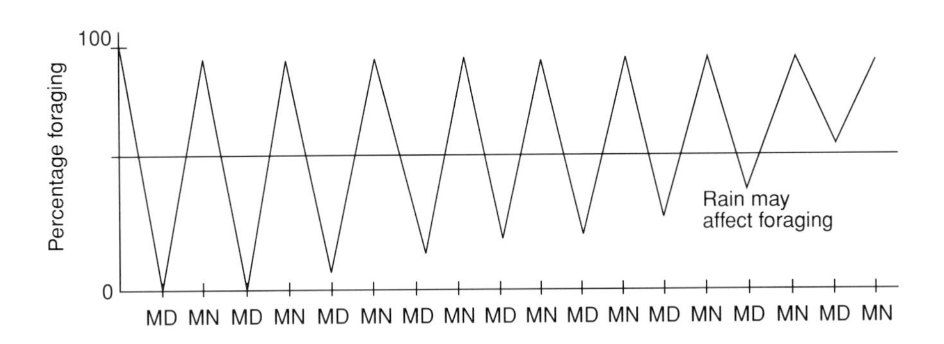

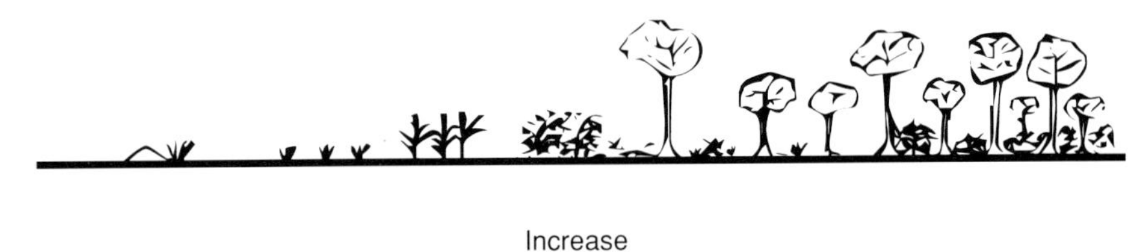

Fig. 43. An overview of the effects of vegetation cover/shelter on the degree of termite foraging.

in turn is greater than farms. Some harvester termites store grass, which helps to insulate them from heat. If they did not store grass, i.e. *Trinervitermes,* they would be confined to more sheltered areas. In sandy regions, the surface temperature can be extremely high and any form of shade or vegetation is important. Most tropical soils at 50 cm depth can lose their diurnal temperature variations, while at 1 m annual variation is usually not more than 10°C.

Small mounds become very hot in dry weather and lose heat more quickly over night. Inside nests, where one area of the nest is in direct sunlight, termites will move to the shaded side. They may also move below the ground, or aggregate in the centre of the nest. For the Australian genus *Amitermes,* which produces wedge-shaped mounds, the flattened sides in the north–south direction give rapid warming in the morning, but the thinner sides expose the smallest surface area to the midday sun, which prevents overheating. However, the same nest orientation has been found in forests, so other reasons for the termites building in this direction may exist. Termites inside the nest of *Amitermes meridionalis* move from the east side in the morning to the west in the afternoon. This is also seen in other termites, such as *Amitermes hastatus* and *Trinervitermes* in southern Africa.

In the Macrotermitinae, metabolism of collected food in fungus-combs produces carbon dioxide, heat and water. This heat can maintain nest temperature. Macrotermitinae fungus-comb chambers remain at 29–30°C in mounds under shade, but within 2–3°C of this in exposed areas. Fungus-comb chamber temperature is higher than in the rest of the mound. Inside wood, temperatures are fairly constant but start to vary as the wood is eaten out and the air flow increases. Termites will then move through the wood to the region of best humidity and temperature.

Subterranean termites tend to burrow near heated parts of buildings and near moist areas. In Nebraska, USA, *Reticulitermes flavipes* can be found under wood in frozen ground and can withstand the presence of ice in extracellular tissues. *Zootermopsis* in Canada can survive at −20°C. Termites survive better in close groups by reducing water loss. *Trinervitermes* moves below ground in the dry season, which often avoids the fires that occur before the rains. Mound-dwelling termites can open and close exit holes to chimneys to help regulate ventilation and temperature. Rainforest termites are not subjected to extreme daytime temperatures.

Food Availability

If there is an even distribution of food in the termite's habitat then there is a definite relationship between nest size and foraging area, i.e. the more termites, the more food required and the larger the area searched. In some termites, such as *Cubitermes fungifaber,* when nests are damaged the termites move to a new site, which also may provide a new food source. Often there are annual cycles in the growth of mounds and amount of foraging. In dry areas, where there is often a shortage of wood, a higher proportion of grass and litter is eaten than in humid areas. Previous cultivation can also affect termite attack, e.g. wheat followed by rice could increase attack.

Stressed plants are also prone to attack. In some areas where soil moisture is low, plants will wilt in the day but recover at night. During the day, when these plants are under stress due to water loss, they are more vulnerable to attack by termites. Crop residues and mulches, such as grass, can attract termite activity. Depending on the timing of the presence of a food source one may be able to distract the termites away from a valuable food source. On the other hand, if food is made available before planting, this could increase the population and therefore numbers that will attack the new crop.

Simple rule-based models using environmental factors can be constructed to predict the probablity of termite attack on certain crops in certain areas, in different soil types and at different times of year. These models examine how situations change with time and other factors and involve if-then-type rules.

The following components can be used to predict the likelihood of foraging and high population numbers in termites:

- Environmental components include month, soil type, rainfall/flooding or irrigation, presence of an available food source, e.g. crop residues, and indigenous food sources such as shrubs.
- Termite components can include an estimate of increase in population size (0–10) based on information from baiting, a change in size (−2 to +2), increase in termites from immigration, e.g. from alates (0–5), and the degree of foraging (0–4).

Predictions can then be made linked to environmental components as to whether foraging and population size will increase or decrease. This model can also be used to include daily/monthly seasonal information on temperature and humidity changes. Assumptions are made from knowledge of the termite species present in the area, e.g. whether they prefer clay soils, or forage more often after rains. Where temperature is included, one has to consider relationships between microclimate changes and high or low temperatures.

PREDATORS AND PARASITES

Termites are commonly caught when they fly, but predators will also raid the nest or catch them while foraging. Termites are an important part of the food chain for many animals such as insects, birds, reptiles and mammals (including man) and they provide essential constituents for a balanced diet (Rajagopal, 1984). In arid regions they may also provide a water source.

Alates

Rains often coincide with the need for food and, because of lowered temperatures and the presence of wet soil for building, termites become more active after rains. Alates can be produced all through the year in lower termites, but flights may only be once or twice a year in higher termites. Flights are often at dusk or in the dark when fewer predators are active. Many millions of alates are eaten by predators as they emerge from their nests. Large numbers are produced at several flight holes, which helps to ensure that some alates escape to establish new colonies. The presence of so many alates means that the predator rapidly becomes gorged and loses interest in eating more termites, enabling some to escape. *Pseudocanthotermes* males flying in trees can attract birds, allowing other pairs to reach the ground.

Alate predators also include red velvet mites in Sudan, which are active after rains, and catch alates that are flying at this time. Dragonflies also catch alates in flight. Birds are the major predators and their distribution can be affected by the distribution of termites and the times of alate production. South African migratory eagles eat *Hodotermes* alates as a major part of their diet. At dusk predators include birds, bats, toads, geckos and ants.

Attraction of alates to light at night means other predators, such as lizards, mantids and spiders, all of which are also attracted to light, are present around the light source. Other predators including ants and small nocturnal mammals and reptiles are also present underneath on the ground.

Large alates (e.g. *Macrotermes* and *Odontotermes* spp.) are commonly caught and eaten by humans or fed to chickens or fish.

Predators and Parasites of Termites in Nests

The presence of ventilation holes or shafts that open into the nest can provide easy access for parasites and predators. Termites have at least twenty kinds of obligate ectoparasitic fungi, e.g. *Termitaria, Antennopsis* and *Laboulbeniopsis. Antennopsis gayi* in Sulawesi (Celebes) has been found on all parts of the cuticle of several pest species of termite (Pearce, 1987). This may have an effect on population numbers. It is often advantageous for the fungus not to kill the termite, but some fungi, such as *Aspergillus,* can kill rapidly. Predators can live inside nests, in tunnels or ventilation shafts. Land planaria can prey on termites, e.g. African *Odontotermes,* by waiting at the nest opening. Mites, e.g. *Acotyledon formosani,* attach themselves to the bodies of termites and interfere with movement and feeding.

Ants will attack colonies. Some *Campanotus* species live with termites such as *Amitermes* species, and will occasionally feed on them, but they

Fig. 44. A ponerine ant attacking a termite of the genus *Macrotermes.*

will also defend the majority of the colony from other ants. Other ant species raid nests or break into them. Raiders of nests include the dorylines and ponerines (family Formicidae) (Fig. 44). Doryline ants can cause very high mortality in *Macrotermes* nests in Africa, often killing the whole nest. The *Megaponera* ant species will respond to soil runways of *Macrotermes subhyalinus.* Ponerines can remove many tens of thousands of termites per day from large nests, e.g. *Macrotermes,* but this is compensated for by a high rate of egg production. Young colonies are therefore more vulnerable to serious losses. Predators such as ants can be sealed off or buried with the same soil that is used for building.

Large predators, such as aardvark and anteaters, break into mounds or wood using their large claws. Their long tongues penetrate into deep cavities. An aardvark can detect the warmer side of a mound where the termites are and dig on this side. South American anteaters and pangolins will climb trees to reach termite nests. The giant anteater of Brazil eats at least eight species of termites, including *Syntermes, Cornitermes, Coptotermes* and *Nasutitermes* as a main food source. It also eats soil with the termites. Armadillos in Brazil eat many *Cornitermes* and *Ruptitermes,* as well as other insects.

Some termites are repellent to predators, especially if eaten in large numbers. The aardwolf, which does not break into the nest but eats foragers, can withstand these and feeds almost exclusively on *Trinervitermes* (Kruuk and Sands, 1972). Echidnas from western Australia reject *Drepanotermes* and *Nasutitermes,* because of secretions, but eat *Trinervitermes.* They also prefer woodland areas, as these areas are where the termites are abundant. The echidnas' sublingual glands produce a sticky substance that causes termites to adhere to its tongue. Sloth bears from India and the USA dig out nests that have not yet fully formed into mounds after the showers start.

In Africa, chimpanzees use sticks or pieces of grass or leaf stalks to remove termites from holes in nests. Mandrills from Cameroon also eat termites. Lowland gorillas in north-east Gabon eat *Cubitermes sulcifrons* and *Macrotermes*.

Predators of Foraging Termites

As well as ants, foraging termites can be attacked by parasitic diptera (scuttle flies). Scuttleflies can lure termites away from the colony and oviposit in their abdomens. The sarcophagid fly from Botswana, Malawi, Namibia, South Africa and Tanzania is a parasite of *Hodotermes mossambicus* in these countries. Parasites can be found inside the heads of soldiers. This causes an increase in size and changes the shape of the head.

Reduviid assassin bugs capture workers of *Nasutitermes* and dangle them down into the chambers to attract other termites that they then catch. A beetle larva, *Pyreearius termitilluminaria* (Elateridae), lives inside *Cornitermes* mounds. It produces a light at night by luminescence and attracts termite alates which are caught and eaten.

Lizards are the major predators of termites in arid and semi-arid regions and termites can provide an important water source for them. The terpenoid chemicals in the defence secretions of some of the termites of the subfamily Nasutitermitinae (Fig. 32; Table 3) are repellent to many lizards. Termites are an important food source for many amphibians that are active particularly after rains when termites are often more abundant. Bat-eared foxes in Africa are estimated each to eat over a million *Hodotermes* per year.

Termites as Food for Man and Domestic Animals

Termites are eaten in many parts of the world. Logan (1992) has reviewed the importance of termites as a food source in Africa. Termites are an important source of protein and may be the only source in some areas, especially at times when little other food is available (Fig. 45). Mounds are of great value in some societies and in some countries may be owned by certain members of a family group and handed down to the next generation. Calories 100 g^{-1} can exceed that of fish, meat, cheese and groundnuts. Percentage protein in some *Macrotermes* alates can be twice that present in groundnuts. In some *Odontotermes* and *Macrotermes* alates nearly half of the insect is fat, the other half protein. In *Hodotermes* 60% of the dry mass is fat. Alates are also a source of many of the essential amino acids needed by humans. They can be eaten raw, dried or roasted in oil and are often found in markets where the wings have been shaken or burnt off after roasting. In Africa termites can be sun dried and ground to a powder and added to food as a paste,

The queen, especially of the fungus-growing termites, such as *Macrotermes* and *Odontotermes,* is considered a delicacy and is also said by some to be an aphrodisiac. In the Congo they are eaten only by women.

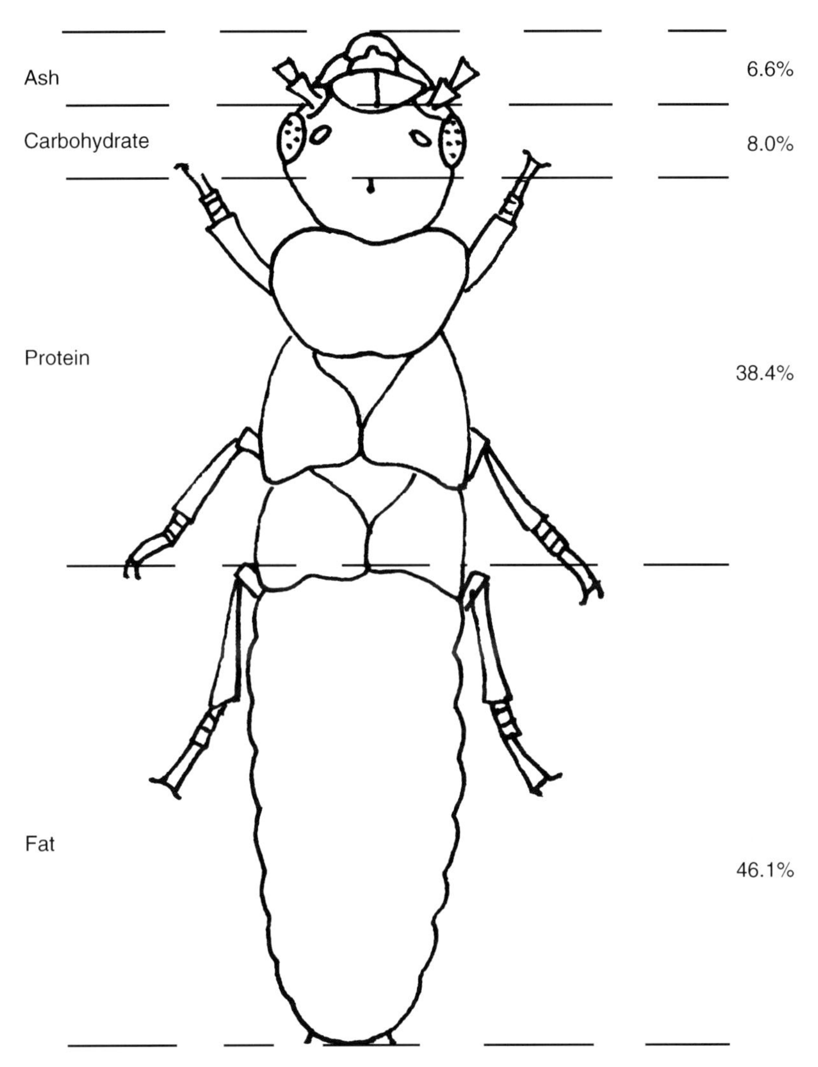

Fig. 45. The nutritional value of an alate of *Macrotermes subhyalinus*. Alates also contain many minerals, fatty acids and vitamins and 20 amino acids (after Oliveira *et al.*, 1976, in Logan, 1992).

When collecting queen termites, some farmers in Uganda know exactly on which side of young mounds the queen cell can be found by the direction of the full moon.

By blocking up most of the exit holes and leaving a few open, the emerging alates are captured. One method is to use skins, cloth or leaves to cover a frame of branches that covers the remaining exit, but under which a small pit has been dug to collect the alates (Plate 15). Methods used to entice alates out of the nest include adding water to the soil, drumming on the mound to simulate rain or using torches at night to attract them. Grasses are also used, in the same way that chimpanzees use them in Africa, by the Amazonian Indians to pull termites out from mounds. In America, there has been a case where a woman was taken to hospital with stomach cramps as a result of accidentally eating live termites, which were able to survive in her stomach (Connors, 1963).

Termites can be used as a food for pigs, poultry and fish. In Togo it is common for pots full of plant waste to be placed on top of *Macrotermes* nests for 3–4 weeks, after which they are removed and the captured termites fed to young chickens and guineafowl. In India, swarming alates may be given to chicks of jungle fowl and in Africa they have been used in ostrich farms.

Australian Aborigines use termites as an antidiarrhoeal agent and tunnelled wood for didjeridoos. In South America, the smoke from burning carton nest material has been used previously for curing ailments of the chest and joints and is still used to repel mosquitoes.

In China, as in some other countries, the mushrooms growing out of termite mounds (as a result of the fungus gardens) are eaten. Extracts of termite faeces and fungus gardens are used as remedies for liver disease and tumours in China. The fungi themselves are often sold in markets in Asia and tropical Africa.

OTHER USES OF TERMITES

Mound Material

The high clay content of Macrotermitinae mounds makes them useful for making bricks for buildings and also for floors and roads. Mound soil is also used for making pots, plastering walls, making ovens and is spread out for growing plants.

Termites and Refuse Cycling

Termites may be important in reclamation of landfill sites, and areas where large amounts of rubbish have been deposited. In tropical countries, termites can remove unwanted litter and animal waste, and these may be their sole source of food. Termites often seek undigested fibres in dung, which they then replace with soil or carton.

Chapter 6

Termites as Pests

Termites are known to infest buildings in the tropics, but in their quest for cellulose they may also cause significant damage to crops and trees. They become pests only when their natural habitat is altered in some way by humans. A typical example is with *Coptotermes* in Malaysian rainforest, where the population is normally kept down by predators. Clearing forest for crops or buildings often leaves rotting vegetation and wood (e.g. tree stumps), which provides an initial source of food for termites in newly established colonies for the first year or so. However, the termites must eventually search for new food (timber or crops) and can then become a pest.

Of the few thousand species of termites that exist, only a few hundred cause damage and only around fifty species are serious pests. Not all timber, crops and trees are susceptible, and their resistance may vary with time or stage of growth. Figures 19–23 show the distribution of some of the major pest species of termites.

FOOD PREFERENCES

Wood, plants, grasses, roots, leaf-litter and humus consist mainly of cellulose, hemicellulose, lignin or other derivatives. Termite-colony success depends on an adaptability to eat different foods. Three kinds of termite feeders exist: these are xylophagous (wood-eaters); geophagous, humivorous feeders (eating soil or organic matter, e.g. leaf-litter); and harvesters (grass-feeders). The grass-feeding termites include *Hodotermes, Anacanthotermes, Drepanotermes, Trinervitermes, Nasutitermes* and *Syntermes*. *Macrotermes* spp. also eat grass but some species will eat other things, including crops in the dry season. Humus-feeders are not so selective.

A good review of the feeding habits of termites, including food selection and consumption rates for different termites in the laboratory and the field, is given in Wood (1977).

Wood Preference

Preferences and resistance will vary with the hardness, lignin content or chemical constituents of the wood. Many researchers have examined the effect of wood extracts on the behaviour of termites. Some areas of wood may not be eaten due to their hardness. Hardness is less important in some species, such as *Cryptotermes* and especially *Neotermes,* which can eat one of the hardest woods, teak. Drywood termites tend to have definite wood preferences. Some termites prefer fast-growing tissues, e.g. spring growth, which produces large cells with thin walls and fewer fibres.

The arrangement of teeth within the mandible, i.e. cutting and grinding edges, may influence feeding success. Major workers have a better cutting edge than minor workers. Soil-feeders have areas of their mandibles developed for crushing. Plants with high silica content may cause mandible wear. The presence of organic chemicals, e.g. phenols, guinones, terpenoids, and high concentration of lignins may also affect the areas where feeding takes place. The pH of wood content might also be important. Sapwood, which has more starch and sugar, is generally preferred to heartwood. *Trinervitermes* have been known to eat wood when grass availability was reduced.

It has been suggested that some of the common tree species seen today in some regions have resulted from selection by termites, i.e. the more susceptible ones have been eaten to extinction many years ago. Many of these indigenous trees are therefore more resistant to termite attack and have developed chemical defences to protect themselves. These chemical defences may be present to a greater level in immature trees and crops, making these even less susceptible. The chemical concentrations in trees can vary from the outside to the inside. Older trees may develop cracks in the bark, and the resistant chemicals may not reside near the outer layers of the tree potentially allowing termite attack to occur. Then, once the termite has entered the tree, adjacent living cells may be exposed and the resistance may fall as the cells die and the wood water content is altered. Common termite-resistant trees include *Cassia,* and some *Grevillea, Markhamia* and *Terminalia* spp. Often gums are produced for defence when trees are damaged, as found in some *Acacia* species – this prevents further termite attack.

Salt-tolerant shrubs, such as species of *Sueda* and *Halopeltis,* can also be resistant. In South Africa, mango, citrus and avocado are said to be resistant to most termite species found; however, in Hawaii, and other parts of the tropics, these are susceptible, attack being caused by *Coptotermes formosanus.*

Resistance also depends on choice. In the case of a termite living in a single piece of wood in a door frame there may be little choice, since there is only one food source available. The termite will have to eat this or die. This may also be seen in the field, for example *Gnathamitermes tubifor-*

mans will eat the creosote bush *Larrea tridentata,* which would be avoided if other food sources were available.

Terpenoids in woods have been known to protect timber from termite attack for many thousands of years in some cases, such as for timber found in ancient Japanese pagodas. In the Macrotermitinae, food preference may vary with season but fungus-combs can act as an off-season food source. Fungal growth on wood can be attractive, some termites preferring this to sound wood. An example of this is the attraction of some *Reticulitermes* species to wood infected by *Gleophyllum trabeum.* Lenz (1994) summarizes the food resources needed for growth and caste development. Response by termites to decayed wood depends on the wood species, type of fungus and the amount of decay (Amburgey and Smythe, 1977)

Crop Preferences

Chemicals present in plants can also be phagostimulatory, i.e. cause an increase in feeding. Roots of plants may give out attractive chemicals. Sugar has been shown to be stimulatory (Abushama and Kambal, 1977a), a 1% glucose solution being preferred by *Zootermopsis.* Sugar cane susceptibility is related to size of seed piece and sugar content (Plate 16). This increases with the size of the plant. Plants such as maize and sugar cane, unlike many temperate plants, are able to photosynthesize at low carbon dioxide levels and high light intensities. This enables them to make more glucose per unit leaf area and grow much more quickly with large cells. Both these factors make them more attractive to termites. Also water content of plants is especially important in drier areas. In parts of Africa indigenous plants such as sorghum and pearl millet, cowpea, arabica coffee and teff have some resistance to termites, but records show attack does occur but often depends on the age of the plants. Younger seedlings can contain repellent compounds such as phenols and cyanides. Resistant varieties of groundnuts have been found in India.

DAMAGE RECOGNITION AND DETECTION

Timber

The form of termite tunnelling within wood may be characteristic for different termites. Galleries of *Kalotermes* or *Cryptotermes* (Kalotermitidae) have clean smooth (evenly cropped) walls, narrow runways and connecting chambers containing seed-like pellets, which are thrown out of exit holes. The pellets may have characteristic sizes or shapes for the different genera and species. Dampwood termites tend to have chambers with rough walls and spotted with faecal material. Subterranean termites (e.g. Macro-

Fig. 46. A wooden block totally consumed by termites (top) and replaced with soil.

termitinae), which, of all the termites, can cause the most damage, fill wood with soil (Fig. 46). *Coptotermes* uses carton rather than soil. *Coptotermes* feeding rate is greater than that of the other main subterranean pests, such as *Reticulitermes* (Grace, 1992). This means that *Coptotermes* can cause more damage. Su and Scheffrahn (1987) found that one colony of *Coptotermes* in the USA could produce over 60,000 night-flying alates and therefore pose a major pest problem if they become established and form colonies.

Mastotermes in Australia also can consume large amounts of wood causing serious damage. Wooden telephone or electricity poles are often attacked by termites (Plate 17), as are wooden railway sleepers. Treating wood can prevent this occurring (Plate 18).

Buildings

Some areas within cities tend to be more prone to termite attack than others, so it is important never to reuse timber that could possibly contain termites. Clearing land for buildings or cultivation deprives termites of their natural food source, so that subterranean pest species will survive better than non-pest species. This is one reason why *Coptotermes,* originally a damp forest genus, has become a serious pest of houses (Plate 19) with wide distribution (Fig. 23). Initially after clearing, damage at the new site will increase. The amount of damage will then fall after a while, eventually building up again to pest proportions depending on the amount of new food source available. External wood in houses is less protected after weathering (Fig. 47).

Fig. 47. Damage to a door caused by the termite *Coptotermes* in India.

For local housing using mud, termites can tunnel up through these structures and eat wooden roof supports or thatching. Thatching in African houses can be expected to last 5–6 years. Wooden or bamboo poles incorporated in the mud or cement walls of traditional houses offer an easy means of entry to termites from floor level to other areas. Termites are also known to tunnel through mortar between bricks. In all kinds of buildings good design is important, but this may still not stop termites invading.

Areas of vegetation in close proximity to cities and towns provide a constant source of alates of building pests. One such case is the mangrove swamps in coastal regions of Kenya, which are a source of the drywood termite *Cryptotermes dudleyi.* Also, Malaysian *Coptotermes* are more abundant in outlying habitats, particularly mangrove, which act as storehouses for termites (Salik and Tho, 1984)

Temperature and air humidity are the main factors affecting termites in buildings. The equilibrium moisture content of wood is affected by temperature and water vapour in the air. Williams (1973) produced predictions for the drywood termite hazard to tropical and subtropical buildings in coastal regions of Africa. He found that sea fogs and early morning dews in semi-desert coastal regions provide water for building soil runways. As the temperature is often low at this time, termites can even forage on the outside of the runways.

Other Damage

Termites often burrow through non-cellulose materials that lie in their path. Valuable books can be damaged by termites (Plate 20) as well as cotton material (Plate 21).

Termites in China are important pests of dikes and dams. In South Africa, *Hodotermes mossambicus* builds nests in dam walls, where aardvarks cause damage by trying to dig them out. In some countries, termites attack wooden railway sleepers and even coaches.

Plastics are often eaten by termites. This causes leakages in plastic pipes and power cuts in cables. Termites in dry regions are known to live inside cables as these provide shelter, higher humidity and protection against sudden flooding during heavy rain storms. High-density polythene and hard PVC is resistant to many termites, while polythene, soft PVC, polystyrene and polyurethane foams are not resistant.

In some countries, burial sites can be damaged by termite activity, soil profiles can be changed and even bone can be eaten by some species.

Crops

Termites can cause direct physical damage often affecting the structural support of crop plants (Fig. 48; Plate 22). Extensive reviews of crop damage are given in Harris (1969) for various countries and Rajagopal (1987) specifically for the Indian region. They can also cause indirect damage by interfering with the food crops and water supply, causing the eventual death of part, or all, of the plant. Attack on perennial crops is more serious than on annual crops, as more investment and time has been put in. Plants under stress, such as in drought conditions and those near ripening stage, are most vulnerable to termite attack. Transplant shock may also affect plant stress. Weeding and clearing can cause some stress to surrounding plants by increasing water loss or causing damage. Another effect of damage is to allow the entry of fungi and other pathogens. Attack on crops by other pests also causes stress and can allow the entry of termites. The most economically important termite genera for crop attack in Africa are *Microtermes, Odontotermes, Macrotermes* and *Trinervitermes*. The Macrotermitinae are also important in India and Pakistan, as are some of the Kalotermitidae, e.g. *Neotermes* and *Postelectrotermes*. In Asia, America and Australia, the Kalotermitidae, and especially members of the Rhinotermitidae, are important. Australia also has *Mastotermes* (Mastotermitidae) (Fig. 23), which is also a major pest of crops and forestry.

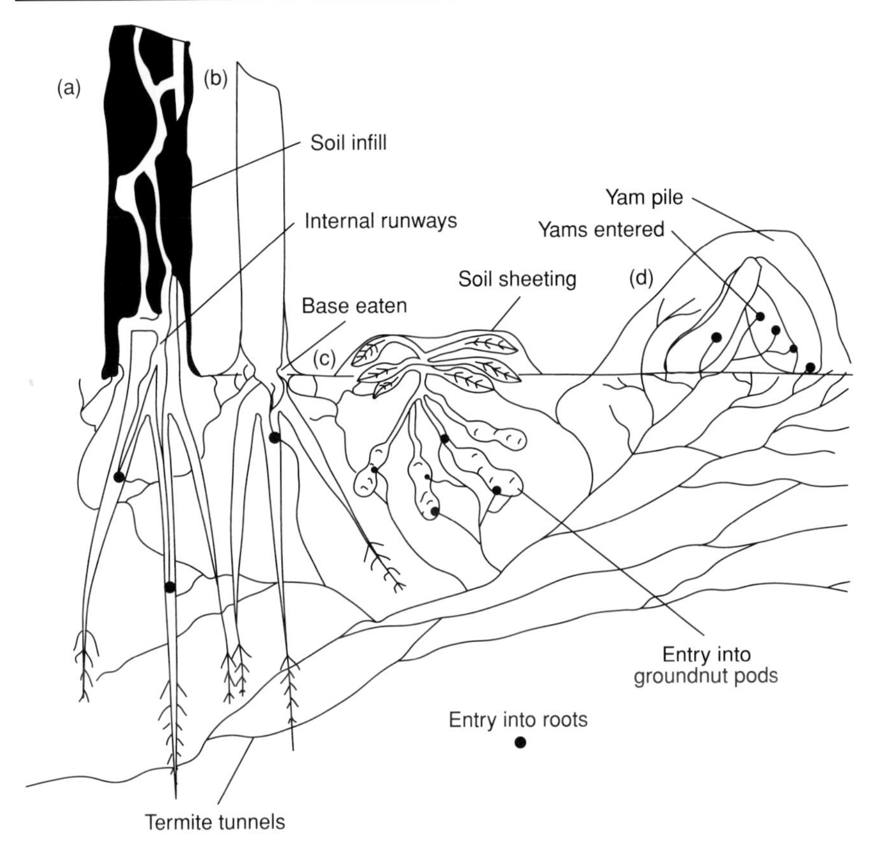

Fig. 48. Methods of attack on different crops by termites: (a) entry through roots, stem eaten out and soil filled; (b) base of plant eaten away under the cover of soil sheeting; (c) a groundnut plant covered with termite soil and pods penetrated; (d) termites entering a yam pile and penetrating the tubers.

Termites can defoliate plants close to the ground and tunnel into, or eat, stems and roots (Fig. 48). Of all the crops attacked, sugar cane is the most susceptible in the tropics (Abushama and Kambal, 1977a; Akhtar and Shahid, 1990). Maize plants fall as roots and proproots are eaten. Wheat is also commonly attacked (Akhtar and Shahid, 1993). Groundnut foliage, stems, haulms, roots and pods can also be attacked by termites (Johnson *et al.*, 1981a) (Fig. 48). Pods are scarified, hollowed out and filled with soil. Damage to pods by termites causes fungal contamination with aflatoxin. This is poisonous and a carcinogen (causing cancer) and therefore reduces the market price. Whole shipments have been known to be rejected because of this.

Microtermes and *Ancistrotermes* species commonly tunnel up through roots or directly into the stem at ground level (Fig. 49). Roots are either completely eaten out or tunnels are made into the stem. The plants will survive for some time as long as some connections (xylem vessels) remain

Fig. 49. The roots of cotton plants penetrated by *Microtermes* in Sudan.

to supply water and nutrients. For *Microtermes* the presence of a layer of soil around the outside of the plant also helps to maintain the humidity and provide support (Fig. 48). *Microtermes* have even been found inside stems and roots of rice growing in water.

Termites, such as *Amitermes,* can attack underground or soil-covered yam sets and tubers (Plate 23). Crops stored in houses or on the ground can also be damaged by termites. Mounds can affect the herding and enclosure of cattle and need to be moved if mechanized harvesting is to take place. Where overgrazing has occurred, especially during dry periods, grass-feeding harvester termites can compete with cattle for food. *Hodotermes* in South Africa collects grass in dry winters, which is a time when cattle may also need food. Loss of pasture, due to dead roots as a result of termite feeding, can lead to soil erosion that will affect soil fertility.

Not all damage to crops is serious, especially if it is close to harvest, where the farmer can rescue the fallen crop, e.g. cobs on maize. Generally, damage is very patchy within fields, but is more widespread in fields containing dry soils.

Trees

The critical period to prevent termite attack in young plantations of tree seedlings is during the first year in the nursery and a few months after planting out. Later on, as the trees get older, termite infestation may increase, but once a good canopy is formed attack is often greatly reduced. A positive correlation exists between the presence of drought conditions and incidence of attack. High demand for timber in Africa has led to fast-growing trees, such as *Eucalyptus,* being planted in poor soil areas where they are under stress and are therefore more susceptible to termites.

Careful pruning and sealing the wounds of trees and shrubs is important to stop entry of secondary pests. Wind damage by rubbing or abrasive action may also encourage termite attack. The presence of neglected plantations close to new plantations is a source of infection. Dead logs on the ground can be colonized by termites. Also dead parts of trees are a means by which termites have ready access to adjacent living parts of the tree.

Coptotermes, Porotermes, and *Neotermes* can completely hollow out trees. Other species including *Odontotermes* also can cause similar damage (Plate 24). As well as internal damage to stems and roots of trees some species of termite will eat around the base of the plant close to soil level. This is commonly carried out by the Macrotermitinae, e.g. *Macrotermes* and *Odontotermes.* This ring barking can then extend up the stem severely weakening the tree and causing water loss (Plate 25). In coconut plantations, galleries in growing tissues can lead to bacterial invasion, which will then cause death. Harris (1969) in his publication *Termites as Pests of Crops and Trees* gives a good review of the kinds of damage to trees and the termites responsible.

Detection

Drywood termites (family Kalotermitidae) need to remove their faecal pellets to the outside in order to extend their galleries and maintain a constant environment. These appear as small piles of small seed-like pellets on the floor, or caught in spiders webs if falling from a ceiling or roof. The area from which they fell should be examined to find the holes from which they were dropped. The wood can be tapped or probed with a screw driver or knife to test whether cavities exist under the surface (Fig. 50).

Subterranean termites are commonly found in damp areas, both inside

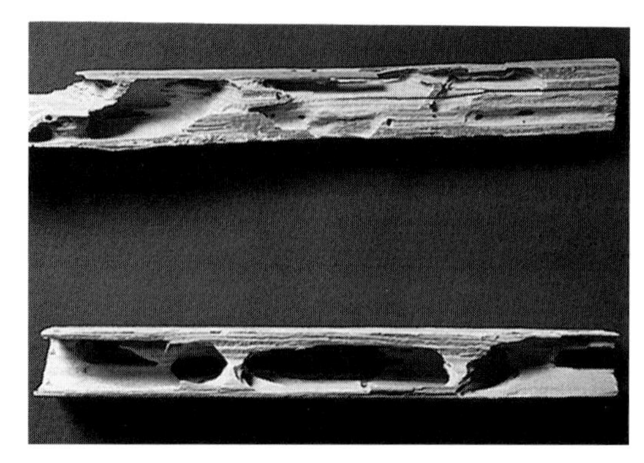

Fig. 50. Wooden shelving tunnelled out by drywood termites.

and outside houses. The presence of soil accumulation or runways is a good indication of their presence.

Another indication that termites have entered a building and possibly entered structural timber is the presence of discarded wings left by the alates. Large wings, over a centimetre in length, probably indicate soil-dwelling termites, whose alates will not form a colony in timber. However, they may build nests outside, or even under, the house and the workers can damage timber. Identification of the genus present can be made from the wings. The alates may have been attracted to lights in or around the house. Several companies in the USA use sniffer dogs to detect termites. Beagles are especially good for this and can go into small areas inaccessible to man.

UV light and x-rays have been used to detect the position of termites inside timber. In laboratory experiments, x-rays have been used to study termites, which has been useful in the study of colony development. The sound made by feeding termites (which is said to resemble typing on a typewriter) can also be detected electronically or with a stethoscope. For *Coptotermes formosanus* a 150 kHz sensor for 70 dB and 0.1 V threshold has been used.

DAMAGE ASSESSMENT

For damage to crops and trees the first sign of termite presence, especially in seedlings, is wilting. Also, soil runways or sheeting on the soil surface, plants or trees indicates that termites are present. Termites can be collected from the runways and identified. To check for the presence of termites in the field, samples of unhealthy looking plants from the crop should be taken. Sampling dead plants cannot rule out that termite attack happened after death, so living plants must also be sampled. Methods of plant mortality assessment are given in Appendix 3.

Chapter 7

Control Methods

In the past organochlorine insecticides were very effective for the control of termites and other insects. These were often applied at higher rates than required to control termites and their breakdown products were also toxic One application around a building could prevent termite attack for over 30 years. But this persistence created potential environmental problems, the chemicals entering food chains and finally reaching humans.

In the 1950s and 1960s the public began to become aware of the long-term effect of organochlorine insecticides, especially after the publication of *Silent Spring* by Rachel Carson in 1962. In 1970 the US Environmental Protection Agency was set up and by 1972 DDT was banned. Other organochlorines then were eventually banned from use in the USA. Today, the main concern is over which chemicals to use. Often new or alternative chemicals are expensive, so farmers and pest control officers, especially in developing countries, sometimes use any chemical that is readily available on the market.

We are now entering a new era of termite control where the emphasis is more on integrated pest management. This involves looking at the ecology and behaviour of the insect and using safer biological and cultural, as well as chemical, methods to alter the pest potential of the total colony. Many governments with environmental organizations, especially in developed countries, have produced initiatives to reduce pesticide use.

CHEMICAL CONTROL

Crop Protection

Chemicals are applied as a planting hole treatment, broadcast over the whole area or placed around the plants. The effectiveness of using the organochlorine insecticide aldrin as a seed treatment, on seedlings, mature

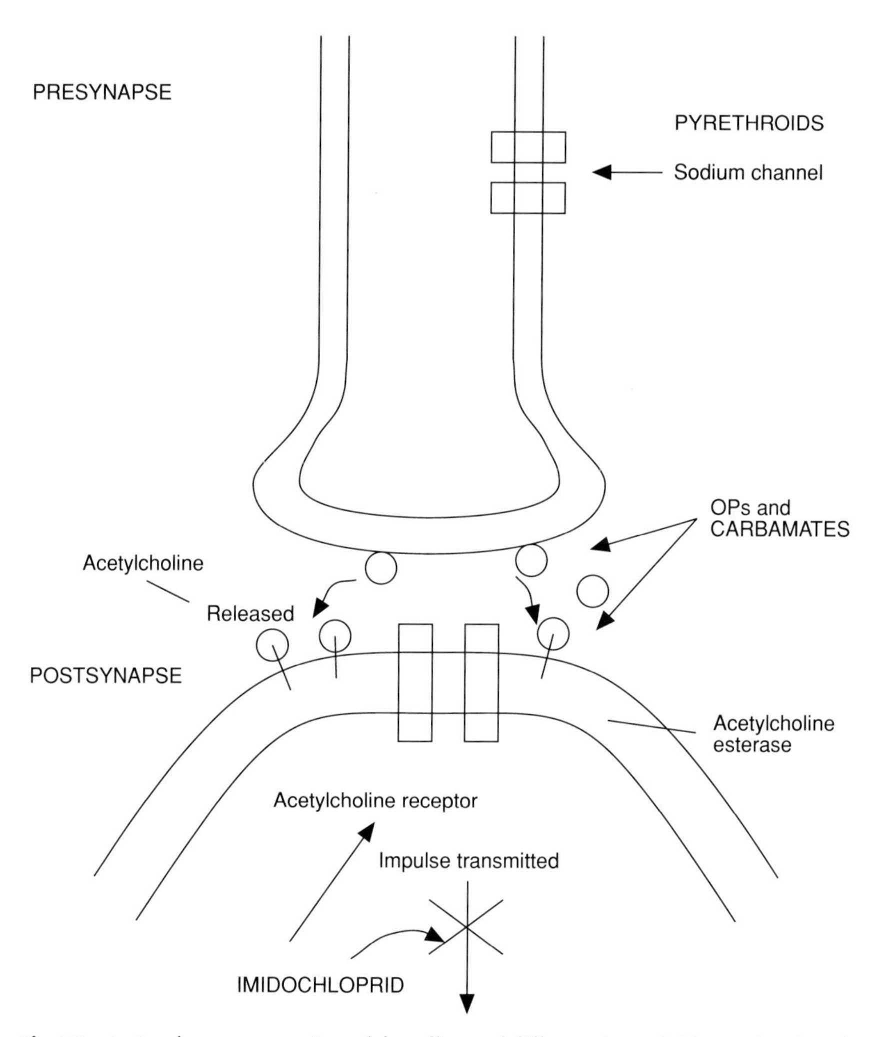

Fig. 51. A simple representation of the effects of different insecticides at the site of the nerve synapse.

plants and for tree protection, is widely known. However, much of this work comes from field trials carried out before the ban on organochlorines. Tshuma *et al.* (1991) summarize some of the methods that can be used today in crop control. Chemicals commercially used in agriculture include oftanol, chlorpyrifos, endosulfan, carbofuran and permethrin. Chlorpyrifos has been widely used for crop protection, especially as a slow-release compound (Plate 26) in crops such as sugar cane, and is effective but costly (Logan *et al.*, 1992). Chlorpyrifos at high concentrations when applied around the stem of cotton plants at 1600 g of active ingredient (ai) per acre can be as effective as aldrin at 400 g ai acre^{-1} (Akhtar and Shahid, 1991).

Endosulfan has been shown to reduce termite numbers for different crops. One of the newer chemicals is imidaclorprid (a nitroguanidine; Bayer) which is a chlorinated nicotine-like molecule similar to nitromethylenes (Storey, 1995). This interferes with the nicotinergic acetylcholine receptors of insects (Fig. 51), changing the social behaviour of the termite, reducing its feeding and resulting in eventual death. It is not repellent to termites, has a low odour and can be also used as a fungicide. Another chemical under investigation is Fipronil (Rhône-Poulenc) which is effective against a wide range of insects including termites. Its mode of action is interference with the passage of chloride ions in the nervous system eventually causing death.

Forestry

Protective methods using hydrophilic grease and filler to prevent entry into pruned or damaged plane trees are commonly suggested as remedial treatments. The major treatments in forestry, as they have been for other crops in the last five years, are chlorpyrifos or carbofuran (Cowie *et al.*, 1989). As forestry includes high-value crops, more emphasis has been put on the use of granular applications, which have a slow-release formulation, especially in nursery seedling application. Release over a two-year period can help seedling establishment, allowing them to become more resistant to termites. A slow-release granule has an active ingredient mixed with an inert ingredient that is thermoplastic and contains many micropores on the outside which release the insecticide. The rate of leaching will depend on soil type, moisture, size of granule and characteristics of the components of the granule. Carbosulfan or carbofuran granules are nearly as effective as chlordane at 0.6 g of ai per tree. Work in Africa, Asia, South America and Australia has shown that the application of Marshall suSCon can provide protection from termites for up to two years.

The pyrethroid tefluthrin and the organophosphate fonofos have been shown to be effective against *Cornitermes* in Brazil, but fonofos at high doses can be phytotoxic to plants. It is important when transplanting seedlings previously grown in treated compost to make sure that the top of the compost is placed above the soil surface when replanting, so that the termites cannot attack the stem above ground.

Rangeland

Examples of grass-feeding termites are *Hodotermes* and *Microhodotermes* in South Africa, *Odontotermes,* and *Trinervitermes* in other parts of Africa, *Anacanthotermes* in the Middle East and Asia and *Cornitermes* in South America. Normally these are not pests but they can become important when cattle are held at high stocking rates. Thus, the cattle eat the grass down to ground level and the termites finish off the remains of the plant including the roots. This can eventually lead to soil erosion and gulleys. Baiting with toxic chemicals has been used to kill grassland termites. Control alone is not sufficient. After control, no livestock should be allowed to enter the area until the grass recovers. Non-fertile soils or over-farmed, exhausted soils which have only poor grass growth are particularly vulnerable.

Nests and mounds in rangeland can cause problems for cultivation. The gas phosphine can be used to control termites (Boicca *et al.*, 1992). Phosphine gas was shown to cause very high mortality of *Cornitermes* in Brazil at the soil surface level, but not below. Methyl bromide gas at 0.45 kg m^3 has been shown to cause total mortality in a few hours for *Macrotermes* in mounds in Kenya. A lower concentration could possibly be just as effective as this can kill the repoductives on which the colony depends. The nest is sealed off during application. Chemicals can also be introduced into mounds through metal tubing inserted into the centre of the nest, which is also an effective application method.

Stored Products

Termites can attack stores, if made of wood, or their contents. Subterranean termites are the major pests. The damage caused by attack depends on the material used to construct the store's base on the ground, as well as that used in the construction of the walls and roof. The less wood used the better. Poisonous soil barriers under the base can be used. Soil insecticides such as chlorpyrifos can provide protection for 1–25 years depending on local climate and soil type. In the tropics it may only be effective for 1–5 years. Oftanol, cypermethrin, permethrin and fenvalerate can be effective for 5–15 years also depending on climate and soil.

Ground pits used for food storage need a lining around the walls. Termite mound soil that has been hardened by firing and waterproofed by a bitumen layer can be used for this. Soil must be treated with insecticide if sacks are to be placed on it. Wood ash can also be used where insecticides are too expensive. Groundnut shells have been successfully used under sacks in west Sudan to prevent termite entry. Where food is placed directly on the ground, and it needs to be protected, less-toxic insecticides, or even wood ash, should be used.

Buildings

An increase in urbanization is happening at a rapid rate throughout the world, except for some parts of Africa and Asia. Edwards and Mill (1986), Mampe (1990, 1997), Creffield (1991) and Berry (1994) give various methods of control that can be divided into preventative and remedial. Wiseman and Eggleton (1994) have summarized the different chemicals available for control of termites in buildings, although new insecticides are regularly introduced. A summary of some control chemicals is given in Table 4.

Soil Application

Before recommending any termiticide it is important to look at previous control using the chemical, the label and data sheet, any information on how applicable it is to the location, soil type, environmental conditions and the proposed means of application.

Houses in rural areas can be destroyed by termites within five or more years, while grain stores can be destroyed in two years. Treatment may not be worthwhile where buildings can be replaced cheaply. Insecticide is added to structures by treating foundations before building or by trenching and then replacing with treated backfill. A modification to this method is to use shallow trenches and to inject insecticide into the base using rods. The base of a building should be made in one complete slab if this technique is to be employed. If it has to be constructed in several pieces, the gaps (expansion joints) can be filled with coal tar or pitch. No pipes or cables (electrical conduit) should pass through the concrete, but where this commonly occurs the pipes or cables should be sealed around with non-shrinking grout, termite barrier sand, fine wire termite-proof mesh or the soil around treated with insecticide and regularly inspected. In the past, insecticide has been mixed with concrete or cement, but this has always proved to be ineffective as termites enter through open joints and cracks rather than directly through the concrete itself.

Stockpiles of aldrin, dieldrin and chlordane still exist in many developing countries, but treatment using these is confined to buildings. Organophosphates, such as chlopyrifos, pyrethroids and cypermethrin, are all effective termiticides and can provide protection for between 1 and 19 years depending on soil type and treatment. Some organophosphates are said to be more effective than pyrethroids, but this is not always the case and is dependent on soil conditions. In tropical regions, organophosphate insecticides have a lower durability than in temperate regions. High clay concentrations in the soil bind chlorpyrifos, so it is less detectable and less toxic to termites, so sandy soils may be better for control, but these absorb emulsions poorly causing a content problem. It is, therefore, important not to assume that a chemical will be equally effective in different areas, regions or countries. Some standardization is therefore needed from evidence collected from trials in various areas. This will allow pest control officers to have some knowledge of what to expect from a certain chemical. An investigation of different termiticides in five soil types and locations in Texas has been carried out by Gold *et al.* (1996).

Foam has been shown to improve termiticide distribution under

Table 4. A summary of chemical and non-chemical control methods with uses and disadvantages.

Chemical/pathogen	Uses	Possible disadvantages
Organochlorines	SB, SC,	All have slow breakdown especially
Aldrin, dieldrin	W, ST	in temperate regions. Harmful to
DDT		humans and environment Withdrawn for use in most countries especially crops
Lindane, heptachlor (endosulfan)		Restricted use Heptachlor low soil persistence
Organophosphates Chlorfenvinphos, chlorpyrifos fenitrothion, fenthion, isofenphos phoxim (Chlorpyrifos can be as slow release capsules or microcapsules in baits) (Phoxim is less stable in wet soil)	SB, SC	Low soil persistence, some phytotoxic at soil surface
Pyrethroids Cypermethrin, fenvalerate permethrin, silafluofen	SB, SC	Easily broken down – can be toxic to fish and bees but low mammalian toxicity
Carbamates Aldicarb, bendiocarb, carbofuran, carbaryl, carbosulfan, propoxur. Carbofuran and carbosulfan produced as slow release compound for forestry and crops. Carbaryl used in baits from grass feeding termites.	SB, SC	Some need moisture for dispersal
Toxic gases Methyl bromide Phosphine Sulphuryl fluoride Chloropicrin, used to protect grain	Fumigants for nest and timber treatment	High human toxicity except nitrogen Chloropicrin phytotoxic Methyl bromide has effect on ozone layer
Others		
Arsenic compounds	W	Very toxic
Borax dust and borates	W, BT	Very soluble – leaches out, phytotoxic
Flurosilicates	BT	
Metal salts and oxides	W	Many toxic – especially mercuric
Silica aerogels	W	
mirex	BT	Use in USA cancelled
Imidachloprid	SB, SC	
fipronil	SB, SC	
Wood ash	SB, SC	

Table 4. *Continued*

Chemical/pathogen	Uses	Possible disadvantages
Plant extractives	W, SC	Unknown effects
Hydrophilic grease	Tree wounds	
Coal tar, pitch	Cement joins, W	Not always long term
Diflubenzuron (Chitin inhibitor)	BT	
Hexaflumuron (Chitin inhibitor)	BT, SB	
Fenoxycarb, methoprene, hydroprene – increase soldier ratio, also stomach poisons	BT, SB	More effective with low soldier population
Bacteria, fungi, nematodes, virus	BT or spray	High numbers needed, viability low, and termite avoidance
Shields, netting Steel mesh, particle sizes	Barriers for new buildings	Often expensive

SB, Soil treatment building; SC, Soil treatment crops and trees; ST, Seed treatment; W, Wood protection; BT, Bait or infective treatment. (Many of the disadvantages can outweigh the expenses incurred without treatment. Many of the toxic effects are often due to solvents used).

concrete (Thomas *et al.*, 1993; Robinson, 1994). A good continuous band of protection was produced by the application of foam through only a few holes. It must be remembered that in very high numbers some termites such as *Coptotermes* can bridge any insecticidal barrier. Termites can pass over a barrier where a water leak is present or where debris has fallen.

Several socio-economic considerations are needed before deciding on using termite control in buildings. These are summarized below:

1. Labour availability and expenditure.
2. Extent of protection provided.
3. Availability of timber, chemical safety equipment and water.
4. Legal requirements.
5. Attitude towards use of pesticides in food storage or schools.
6. Emphasis on traditional methods.
7. Financial status of family.
8. Ownership of the property.
9. Inconvenience of control method used.
10. Availability of outside funding.
11. Status in the community.
12. Degree of damage that has occurred.
13. Importance of building to the community.
14. Presence of incentives and cost of treatment.

Wood Treatment

Traditional methods involve dipping timber in coal tar, petrol, creosote or pentachlorophenol, which can provide timber with protection for a few years. These are often very toxic, have unpleasant odours and discolour wood. Burning the ends of posts in contact with the soil or coating them with engine oil or diesel also provides short-term protection, as do some paints. Common chemical preservatives include creosote, copper chrome arsenate, pentachlorophenol, boric acid and fluoride salts applied as pressure and dip treatments. Metal salts and oxides, such as mercuric chloride, copper sulphate and mercuric, zinc oxides can be effective for 3–5 years. Chlorpyrifos, permethrin, bifenthrin, cypermethrin and deltamethrin are also effective, but more expensive, so their use is restricted to high-value buildings.

For remedial wood treatment (i.e. after a house has been built) infested wood can be removed and toxic or abrasive dusts (e.g. silica aerogel dust, sodium fluosilicate and calcium arsenate) squirted into exposed runways in the wood, which are then covered over with tape. Bendiocarb dust may also be used (Mampe, 1990). The termites pick up and spread these dusts during grooming. Insecticide can also be added to wood by 'drill-and-treat' methods.

Arsenic trioxide or the cyclodiene Mirex have been used to treat entrances and swarm holes in China (Li *et al.*, 1994). It is often better to replace the wood where damage has been caused by drywood termites, as a new colony can be established from only a few workers that may not have been killed by treating.

Treatment with borax is standard in several countries, although in others it is registered as a fire retardant, but not yet as a wood preservative. Borax compounds are cheaper than creosote and less volume is needed to be effective than for copper chrome arsenate. Disodium octaborate tetrahydrate is the most useful as it is water soluble (Jones, 1991; Moore, 1993). Boron-based compounds are highly toxic to wood-destroying insects and also to fungi and have a low mammalian toxicity. An aqueous solution of sodium borate can be used for sprays and for 'drill-and-treat' methods in houses (Plate 27). However, borate-treated wood is not chemically fixed like those treated with copper, chromium or arsenate salts, so near the ground or in wet conditions the borate leaches out. Prevention of leaching can be achieved after treatment by the use of a water repellent. Boron treatment is best if used on unseasoned timbers to allow diffusion through the whole timber via the water already in it. Treatment can be done by dipping. Brushing on a hot solution is effective, but vacuum treatment is better although more expensive. Treatment of wafer board and other composite materials is more effective than treatment of wood, as diffusion is better (Myles, 1994). The leaching of borate from wood in wet conditions can, in some cases, be useful, e.g. at the base of telegraph poles or under floors.

Fumigation as a method of wood treatment is still an important means for termite eradication (Plate 28). Sulphuryl fluoride and methyl bromides in a double-bag enclosure are still recommended. A nylon polymer bag will provide the best protection. The building must be gas-tight before treatment. Sand is used to hold down the plastic sheeting covering the building and safety precautions are applied. For methyl bromide, a concentration of 1.5 kg 100 cm^{-3} for 34 h is needed. Concern has recently been expressed on the effect of methyl bromide on the ozone layer. Alternatives available include sulphuryl fluoride (Dow Chemicals) and carbonyl sulphide (Delate *et al.*, 1995). For museums, libraries and archives, nitrogen or carbon dioxide gas used as an anoxant is said to provide a safe method for eradicating drywood termites. Liquid nitrogen can be used for small items where it is pumped up into walls and voids. Temperatures of −19°C can kill termites inside wood but the edges must also be cooled. Another method of killing termites is using electroguns, which rely on damp galleries to conduct current to where the termites are living so that they die. Another method which is proving promising is placing the termite-infested timber inside tents and using large propane burners to raise the temperature to 66°C for at least 4 h. Termites will die if subjected to temperatures of 55°C for 5 min or 49°C for 30 mins (Woodrow and Grace, 1977). This is useful for spot treatment especially for drywood termites. Careful monitoring of the humidity inside the sealed container and the temperature of the wood is carried out during the process, and thermal treatment also means that houses can be reoccupied sooner than if using fumigation treatment. Cables can be protected by incorporating carbaryl insecticide, where the resistance depends on a bloom layer on the surface.

Baiting

Baiting has the advantage of not contaminating the soil with chemicals. Baits can be used where insecticide treatments are avoided for fear of contamination or where owners cannot afford extensive treatment. One method is to put out non-toxic baits near colonies for the termites to locate and forage and then to replace the baits with toxic ones. The effectiveness of baits depends on the foraging behaviour of the termites (i.e. termites first need to find them), the kind of bait (shape, size, content, etc.), its odour and the way it is presented. Termites may initially forage at random, especially when food sources are low, so baits need to be widespread in order to find out where termite activity is the greatest. More baits can then be placed in this region so that more termites can be recruited to forage. Baits must be attractive, and better still, more attractive than the surrounding food sources. For some species of termites sugar, molasses and honey can be used to increase bait consumption. Other baits may be made attractive by treating with moulds, allowing partial decay, adding amino acids, nitrogen sources and even attractive pheromones held in vegetable oil. As termites, especially in dry areas, are attracted to wet soil, water can be added to the soil under the bait. The bait itself can also be constructed so that it holds moisture.

Over the last five years many potential baits have been tested in the laboratory and found to be toxic to termites. The chemical should ideally be a slow-acting poison that is transferred throughout the colony. It also must be safe to the public and not leach out into the soil. Examples include Mirex on *Mastotermes*, diflubenzuron and hexaflumuron for *Reticulitermes* and *Coptotermes* (Su and Scheffrahn, 1990b, 1996; Su, 1994b; Clement *et al.*, 1996) and sulfluramid (Myles, 1994). These fluorine compounds affect cuticle hardening so that moulting is incomplete and the termites die. Growth-regulating chemicals such as fenoxycarb, methoprene and hydroprene, which cause over-production of soldiers, can also be effective where soldier numbers are usually low (i.e. a greater demand is placed on feeding these and the colony fails). In lower termites these chemicals may kill the gut protozoa, again causing mortality.

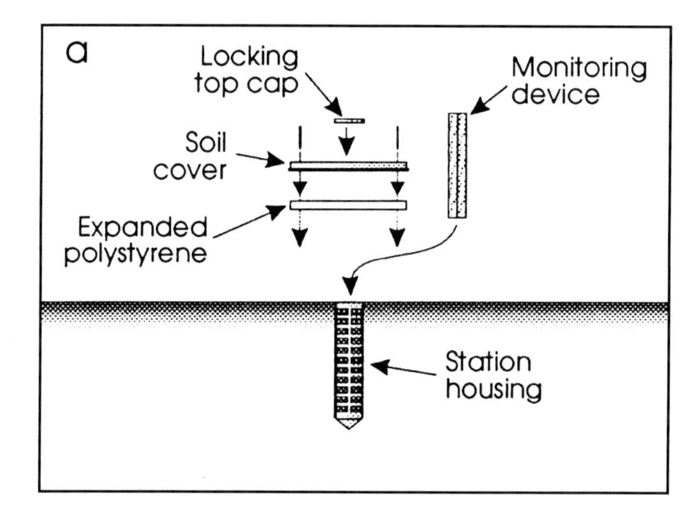

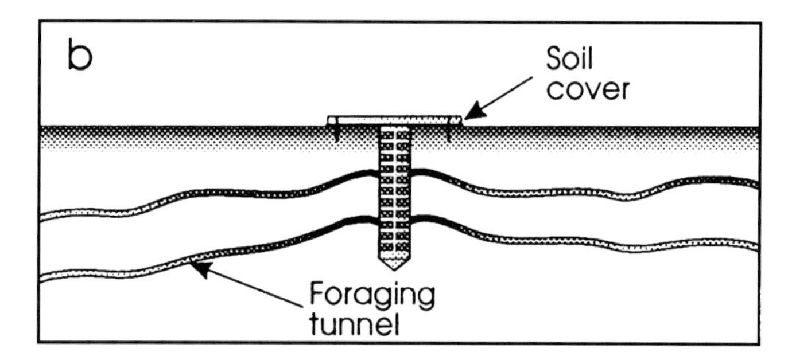

Fig. 52. (*and opposite*) The 'Sentricon' bait station for monitoring termites and feeding slow-acting insect growth regulator (Dow Chemicals).

One method of baiting is to use a plastic conduit placed vertically in the ground. Inside this is a smaller conduit containing corrugated cardboard, which can be removed easily (Plate 29). A commercial variation of this for use around buildings, which has proved successful and incorporates the use of a growth-regulating chemical, has been developed as the 'Sentricon bait' (Dow Chemicals, Fig. 52). A similar subterranean bating system, 'Subterfuge' has been developed by American Cyanamid but bating is not monitored.

A 'trap, treat and release' method was also developed by Myles (1994) for controlling *Reticulitermes* in Canada. Corrugated cardboard is again placed in piping in the soil to attract termites and the collected termites are treated with latex sponge containing sulfluramid and then released. By collecting termites over a few weeks using this method, several hundred thousand termites can be treated and released together. A similar and equally effective method for collecting termites that is employed in Egypt is to use corrugated cardboard pushed into dry soil with a plastic bag over it. For the

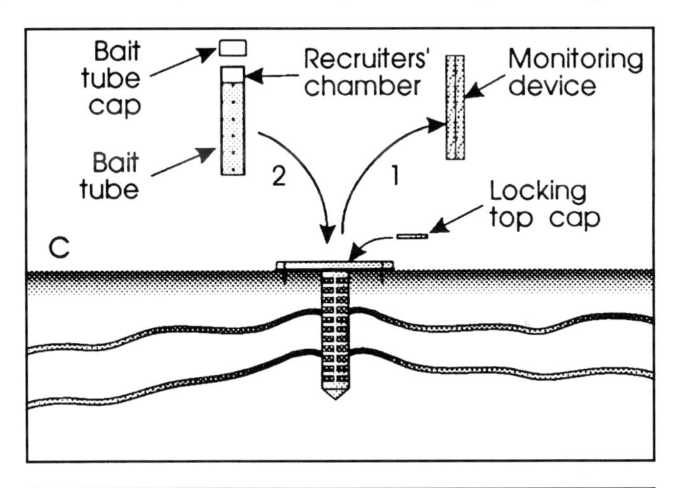

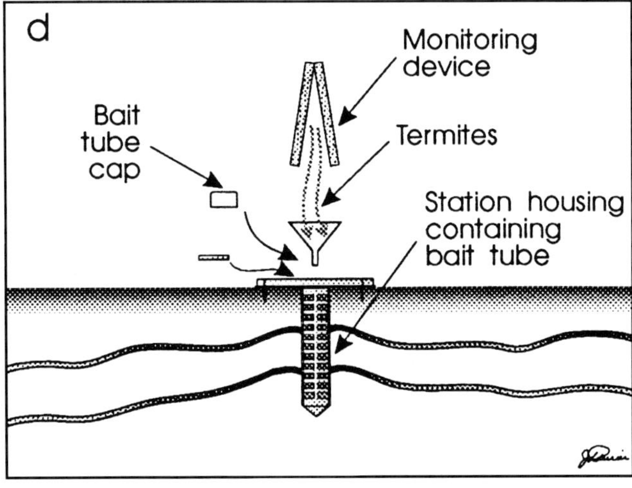

fungus-growing termites, trials have been carried out using fungicide-treated cardboard discs. These are placed out in the field (Plate 30) and consumed by the termites, which then deposit the fungicide on to their fungus-combs on which they depend for survival. This method has been shown to work well in the laboratory (El Bakri *et al.*, 1989b), but not in the field. The main reason for this is that the termites eat other foods as well as the baits, which dilutes the effect of the fungicides. To be successful with this method the site would have to be cleared of vegetation and the area saturated with treated baits, or a bait that is more attractive than the surrounding food sources should be found.

Indoor bait stations ('termite picnic boxes') can be placed next to places with termite activity. This method has been used by CSIRO in Australia (Plate 31). The bait stations are made from polystyrene or wood and have slits in the sides and a lid with a transparent window allowing observation of the termites. The box is filled with layers of wet cardboard as the bait, buried, water added to surrounding soil and the box examined after a week to see if termites are feeding. Tubes can be used to link inaccessible nests inside trees to the box to attract termites. These boxes can be placed inside houses near termite infestations at floor level or below the floor. Control is achieved by adding arsenic trioxide, or other equally effective chemicals, to the termites on the bait in the box so that they carry the poison back to the nest.

Recently the FMC Corporation (US) have developed a bait station at the soil surface ('First Line'). This consists of treated cardboard in a plastic box with the edge of the station at a 45° angle to allow termite entry. These boxes are placed near termite tunnels. Other companies are looking to develop similar baits.

Boric acid and related compounds have become widely used over the last few years, especially as wood preservatives. These have been used against *Coptotermes* and *Reticulitermes,* which are major pests in the USA (Grace, 1991).

Baits used in pasture have included sawdust impregnated with boric acid and a binder containing honey and moulded to give a control agent. Boric acid and a plant-based attractant have been used as pegs placed in floors of houses or in roof areas.

The possibility of using microencapsulated permethrin in baits has been examined by Schoknecht *et al.* (1994). Capsules are spread through grooming and trophylaxis and can enter the gut, remaining intact so that they can be passed on further within the colony before finally having a full effect. The use of attractants, especially phagostimulants, is an area that needs further research. Waller (1996) has looked at the possibility of using urea solutions for controlling termites in Australia.

Baited traps without chemicals have also been used to reduce termite numbers. Pits filled with attractive food or bait boxes are used, the contents of which are removed later with the termites. It is very important to be able to judge the population present in an area and the foraging territories before monitoring the effect of baits (Su and Scheffrahn, 1996). Monitoring also should be carried out away from the toxic bait site and at treated and non-treated baits.

Any new method, such as baits, which will eventually be introduced to the consumer has to be tried by pest control operators. Good training, quality assurance and an effective monitoring method are required throughout the test period (Thoms and Sprenkel, 1996). Unlike other chemical methods, these new methods may need time to be totally effective but they have the added bonus of being localized and contained, and therefore safer to handle and more 'enviromentally friendly' than many insecticides.

Plants and Plant Extracts

Plants or plant extracts can provide a simple means of control that can be used by farmers. Plants can also be used to attract termites away from crops, the leguminous tree *Leucaena* being one example. Some plants may possess antifeedants. These include limonoids from plants in the families *Meliaceae* and *Rutacea*. Tree resins such as sesquiterpenes especially of primary forest trees, are effective termiticides. A document (*sutra*) kept in a wooden pagoda in Japan for over 1000 years was well preserved because the paper was coated with an extract of amurense.

Various trees and leaves in India have been shown to be effective against termites. Examples include deodar tree bark and leaves, chiv trees, castor plants (oil) and neem tree leaves. The commercial preparation of neem, Margosan (0.3% azadiractin with 14% neem oil), at 100 ppm can cause significant mortality of *Coptotermes* but lasts only a short time (Grace and Yates, 1992). Neem mulches are somewhat deterrent to *Coptotermes* but may not be effective for all termites (Delate and Grace, 1995b).

Latex from *Calotropis* can be used, as well as fresh onion for drywood termite control. A cattle melon mixed with water is said to be repellent to termites as are extracts used to kill fish. Black pepper, which is toxic to many insects, may also be used against termites.

Residues and Contamination

Workers from the USA, Australia and Japan have looked at the effects of contamination caused by use of organochlorines and also, in some cases, pyrethroids. They found that using chlordane or hepatachlor, the young and old were exposed to airborne treatment more than humans of other ages, and a considerable amount of residual vapour resulted from past treatments. In Japan, it was shown that chlordane could be found as oxychlordane in the blood and transchlordane was found in the sebum. WHO

(1989) showed the presence of aldrin and dieldrin in mothers' milk and suggested that these organochlorines also caused a higher mortality rate in birds than did the use of DDT. Taguchi and Yakushagi (1988) looked at the effects of termite treatment using chlordane on the concentration of the chemical in human milk and found a continuous accumulation of chlordane in dwellers of treated houses, persisting for at least five years after treatment. Sasaki *et al.* (1991) have also looked at skin lipids as an indication of organochlorine concentration in the human body. For some insecticides the smell of the solvent used to dissolve the chemical may be unpleasant for the homeowner after the house has been treated.

PHYSICAL AND CULTURAL CONTROL

Barriers

A summary of the precautions that can be taken to prevent termite entry to houses is given in Fig. 53. Fly screens can be placed over windows, cavities and vents in areas where drywood termite alates are common. Metal shields of copper or steel or concrete ridges bent downwards at 45° around a building can, to some extent, prevent termite infestation. Shields make termite runways more visible, but regular inspection is needed. However, caps and shields are no longer recommended in some countries as no guarantee can be made for termites not crossing them.

Removal of debris and existing colonies is very important. No extensions should be built in contact with, or material placed next to, affected houses. The lifespan of houses and stores can be increased by raising them on supports of stone or on treated or resistant wood. Chickens can be kept under food stores as they will eat any termites present.

Sand, basalt, granite, glass splinters or globules and fossilized coral of a specific size can be used as barriers (Tamashiro *et al.*, 1991; French, 1993; Grace and Yamamoto, 1993). The grains have to be too big for the termites to move, and spaces between grains too small for termites to tunnel through. Su and Scheffrahn (1992) showed that a range of sand particles of size 2.0–2.8 mm were effective against American species of *Reticulitermes* and *Coptotermes*. A 20 cm thick layer can be placed outside the building or in crawl spaces. Checks need to be made on physical barriers and the barriers must be maintained. Diatomaceous earth may not be an effective barrier (Grace and Yamamoto, 1993). To be successful, all barriers need to be continous and well defined, possibly with slow-acting insecticide on the outside of the barrier. Barriers have yet to be tested in the tropics.

In Australia, a fine mesh of stainless steel placed under a new building has been shown to offer good protection against termites (Lenz and Runko, 1994). Mesh can also be used as a protective measure when wrapped around posts at joints, over cracks in slabs or around electricity cables. A

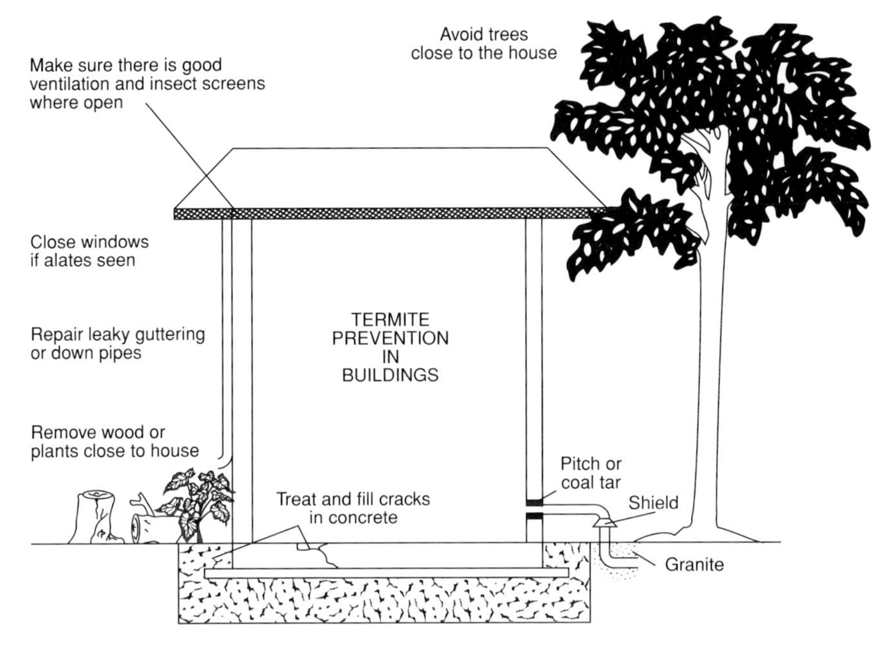

Avoid sloping ground towards house. Chemical barrier or particle barrier such as granite. Check outside areas that stay damp for long periods, e.g. around drains. When buying a house it is important to know past history of termite attack and treatment.

A plastic vapour barrier with a stainless steel wire mesh beneath it can be laid over the foundations when building a house to prevent termite entry. A treated fabric blanket used in the same way has also proved to be successful

Fig. 53. A general summary of the precautions to prevent termite entry to buildings.

modification of this method, used in Japan (Yoshimura and Tsunoda, 1994), is to mix the insecticide with synthetic polymers, and to place one layer of this sheet on the soil with an untreated layer on top. This makes sure that the insecticide does not leak upwards into the house.

Crops

The planting of fodder trees or grass between crops can be used to distract termites away from the main crop at important times of the year. Ash can be placed as a mulch around some crops such as yams in Nigeria. Spacing or chemical feeding of trees can affect the amount of termite attack, so these methods can be adjusted. The first rains can sometimes release nitrogen from the soil, so early planting will give higher crop yields. In crops planted later, the nitrogen is leached and a greater degree of insect attack is common.

Replacement of tropical forest by open forest reduces soil moisture in the upper layers of soil, so that some termites, such as *Macrotermes,* have reduced activity.

Adding water to dry soil can attract termites, but constant irrigation of crops or trees will prevent termite activity and can be used as a control method. Over-watering can cause problems for termites. However, some termites can survive in flooded areas for some time in air pockets in soil or wood. The irrigation scheme in the Geziera (Sudan) has contributed to a reduction of termites in the irrigated cotton area.

Some suggestions for plant protection against termites are as follows:

1. Make sure that you have good quality seed, preferably a resistant variety and practice crop rotation.
2. If possible plant early in a fertile soil so the plant has a good chance to grow.
3. Irrigate regularly and only manure when best for the plants not the termites.
4. Try different densities of plants and a trap crop.
5. Remove diseased or damaged plants and weeds, taking care not to damage the roots of the healthy plants.
6. Dig out or poison nests of pest species.
7. Harvest quickly and remove and burn plant residues.
8. When planting forest seedlings in a nursery, use polythene tubes with treated soil.
9. When planting out, make sure the treated soil remains above soil level and around the stem.
10. Where branches are pruned, make sure the cut ends are sealed to prevent termite entry.

Mound Removal, Flooding and Local Methods

The queen must be dug out of nests to ensure that the colony cannot recover. However, one must be certain there is not more than one pair of reproductives present. *Macrotermes michaelseni* in Kenya has been found to have multiple reproductives (queens and kings) in some of its nests (Darlington, 1985b), so killing one queen may not be sufficient to destroy the colony.

One also needs to ensure that only mounds of definite pest species are killed. Often, all mounds are destroyed in an area even when the main pest species does not build a mound. In Brazil, a mound-drill mounted on a tractor has been developed to destroy *Cornitermes* mounds. For mounds belonging to other species, dynamite and mechanical diggers have been used. Irrigation and flooding of mounds can kill termites. In Sudan, fewer termites are found on well-irrigated trees and crops.

The burying of dead animals such as dogs, goats and other materials

(Omo-Malaka, 1972) can cause termites to move away. In Uganda, dead chameleons placed inside mounds also have the same effect. The cause of this may be that the dead animals attract predatory ants or other termite-eating insects.

Resistant Timbers

Some timbers, but not others, may be resistant to one particular species of termite (Harris, 1971) (Figs 54, 55). Slow-growing resistant timbers are often expensive to buy and may be exported. This leaves the more susceptible fast-growing timbers behind for local use. Resistance depends on many factors such as hardness and chemical content. Wood containing saponins, which are toxic to insects including termites, has been found intact after over 1000 years. Mitchell (1989) tested 41 species of tree from Australia and field trails in Zimbabwe for possible resistance to termites. Of these, several *Acacia, Enterlobium* and *Senna* spp. showed some resistance and were possible candidates for use as timber and in forestry.

BIOLOGICAL CONTROL

Biological control of termites currently involves the use of natural pathogens, that is those commonly associated with the termite, but not

Fig. 54. Japanese Cypress heartwood partly resistant to the termite *Coptotermes.*

Fig. 55. Non-resistant Russian pine attacked by *Coptotermes.*

causing serious harm unless present in large numbers. These include nematodes, fungi and viruses.

Nematodes are effective biocontrol agents for termites living inside mounds or branches, (Danthanarayana and Vitarana, 1987; Logan *et al.*, 1990). Many hundreds of infective nematodes are released from one termite after infection by a single nematode adult (Plate 32). However, the main problem with the use of nematodes for control is the need to bring termites into direct contact with them, and to survive they need free water. Larval stages of many insects, especially beetles, are immobile and so they can be infected fairly easily, but termite larval stages are mobile and can be considered as miniature versions of the adult. Infected termites also tend to be isolated from the colony by the other workers, therefore preventing further infection. Maudlin and Beal (1989) concluded that nematodes are not very effective for subterranean termite control.

The effect of the nematode *Steinernema* on *Reticulitermes* was examined by Epsky and Capinera (1988). This same genus was shown to be effective in China against all castes of *Coptotermes formosanus* and *Reticulitermes speratus* at a dose rate of 4000 to 8000 nematodes in 3 ml. This gave complete control in a tea bush at 40 ml per bush.

The fungal pathogens *Metarhizium anisopliae* and *Beauveria bassiana,* do appear to be more successful for the control of termites. Spores can be blown into nests, held in baits or applied as a wettable powder to the nest or termites. Although termites will isolate, carry away and avoid infected individuals, Fernandes and Alves (1991) using 5 g of isolate succeeded in killing *Cornitermes* colonies occupying nests in Brazil after 10 days. Minor workers are more susceptible than major workers and soldiers. Jones *et al.* (1996) found that fungus isolates differed substantially in their pathogenicity to *Coptotermes*. Isolates from the termites themselves may be the most effective pathogen (Grace and Zoberi, 1992; Delate *et al.*, 1995). Much work on different *Metarhizium* strains has been carried out in Australia by CSIRO, where it has been registered for control (Hanel and Watson, 1983; Staples and Milner, 1996). Australian termites tend to have a more defined nest structure than American termites, so control may be more effective. In the USA a *Metarhizium* preparation is used as a termite repellent (in the form of Bioblast – manfactured by EcoScience) and applied as an inert dust or wettable powder. Eventually, if it can be registered in other countries, its use in the next few years will become more widespread. Other parasitic fungi found on termites may also be useful for termite control by reducing populations (Sands, 1969). One example is *Antennopsis gayi* Buchli (Fig. 56) found on *Coptotermes* in Domoga Bone National Park, Sulawesi (Pearce, 1987).

Al Fazairy and Ahlam (1987) showed that a nuclear polyhedrosis virus isolated from the cotton leaf worm, *Spodoptera littoralis* (noctuid), could infest termites. Further work on the use of viruses for termite control may prove useful for the future, especially as transmission of infection would

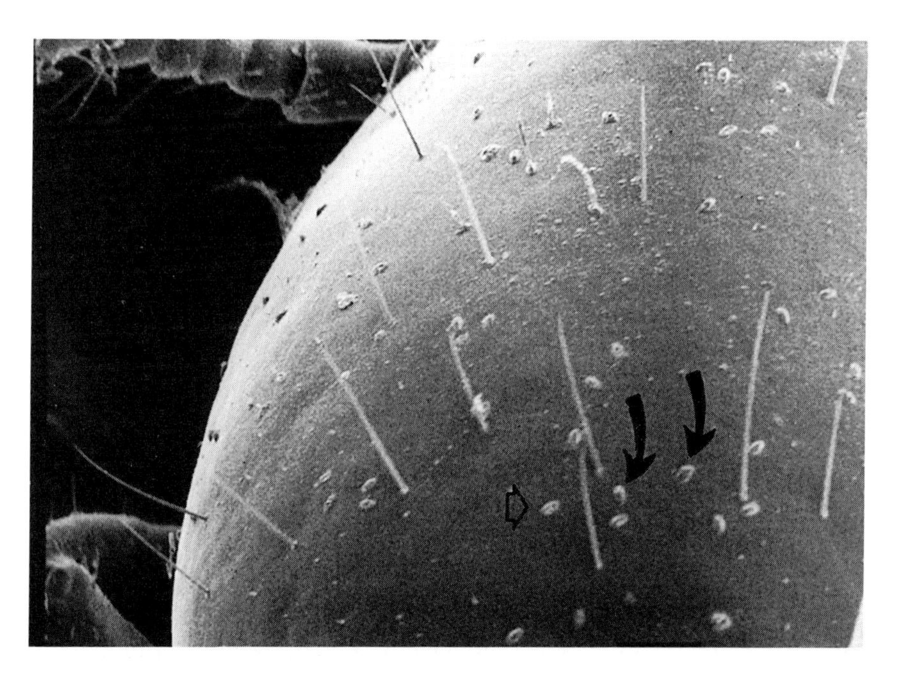

Fig. 56. Haustorial capsules of the parasitic fungus attached to the head of *Coptotermes* found in Sulawesi.

prove easier than for other organisms (such as nematodes). Several applications would probably be needed to be effective. The pathogenicity of the bacterium *Bacillus thurigiensis* in termites has been studied by Khan *et al.* (1985) and Grace and Ewart (1996).

SAFETY

It is now recognized that the bromine in methyl bromide used for fumigation can reach the stratosphere and destroy the ozone layer. Alternatives available include phosphine, carbon dioxide, nitrogen, heat and cold. With the increasing use of microbial pest control a protocol is being developed for the use of microbial pest control agents so as to protect non-target species (Grace, 1994).

Many safety precautions need to be maintained, both during application and after treatment. Persistent insecticides can contaminate soil and water, kill sea creatures and affect the development of birds. The following safety precautions need to be applied:

1. Application should be closely monitored where human contact is most common, for example around a swimming pool.
2. Application and safety instructions on the label must be followed closely. The risk to users from insecticide depends on the toxicity of the chemical and the length of exposure.
3. Full protective clothing is needed (gloves, goggles, appropriate mask for dust and/or droplets, and protective suit). In the tropics this is often not feasible due to high temperatures, but face masks, gloves and some sort of body covering, such as an apron, are essential. It is important to check continually the condition of protective clothing, as it has a limited life. Gloves especially can perish and the protective benefits of masks diminish over time.
4. No smoking or eating is allowed during application.
5. After application or spillage on to the skin, washing with soap and water must be carried out.
6. Proper disposal of empty containers must be carried out as required by law, i.e. containers and chemical (toxic waste) are returned to the supplier, or taken to an approved waste disposal facility (which is clearly labelled as such), where the containers are punctured and crushed and then buried or burnt. On no account should any empty container be used for any other purpose, as remnants of the chemical will remain inside the container even though it appears to be clean.
7. Any small amounts of remaining chemical left after treatment must be returned to the store for future use and not used for purposes other than those dictated on the label.
8. Protective clothing, if not disposable, must have regular washing. Even

then it may continue to build up residues of persistent insecticide that cannot be removed, so it should be destroyed after a period. Clothing that has been used during the application of insecticides must be washed separately from clothes used for normal daily use by the operator or his family, as the insecticide can be transferred to them. Dry cleaning should never be used as this does not remove insecticide residues.

FUTURE CONTROL

Termites are gradually moving northwards in North America, north Asia and northern Europe. Many countries in the world are becoming more developed, and termite pests are becoming an ever-increasing problem.

As urbanization and movement of traffic between countries is increasing, it is becoming more important to check that termites are not transferred. There is a need for more routine examination for termites in soil, plant material and timber by customs officers. Training and guidelines on how to recognize signs of termites need to be increased worldwide.

With increasing human populations, intensive agriculture is needed and this brings with it increasing pest problems. The emphasis on the control of termites in the next few years will no longer be on a single chemical that has all the answers.

An integrated method of control will be adopted where the emphasis will be on several methods. Different methods will be used together or separately to provide a safer, more effective, method for termite control in crops, forestry and buildings. Some methods may be effective on their own or can be used as a backup to other methods.

A thorough investigation of biology and behaviour has not yet been carried out for the majority of termite species. Once this has been achieved, new methods can be devised to manipulate behaviour and monitor population numbers in order to reduce their pest status within a particular area or food type.

Pattern of termite nest distribution is an important indicator of soil health. Any changes will indicate the effects of human activities or climatic change. With the use of new insecticides and methods it is important to have an effective training programme to ensure 100% efficiency and protection. Also it will be important to convince the customer that these new methods are equally as effective as previous methods in the long term. These methods will also have to be investigated in different regions for different termite species, allowing the pest controller to be able to predict the likely outcome of using this method of control. The detection of termites and their distribution will be made easier in the future with the aid of electronic detectors and other marking techniques.

In the future, surface and subterranean baiting could become commonplace, not only for the control of termites in buildings but also in other situations, e.g. crops and forestry. Bait stations containing attractive control agents could be placed in farmers' fields, and termite foraging would become concentrated there. Different baits will have to be developed for different species of termite as food preferences differ. More emphasis could also be put on reducing populations of pest termites from sources surrounding urban areas. A great many pest species exist in mangrove areas and other areas around towns and the constant influx of termites from these areas is a major source of reinfection. Overall, there will be less emphasis on the eradication of termites but more on the manipulation of colonies to create levels that will not cause damage.

The use of termites commercially, as a food source or as a means of breaking down waste cellulose (e.g. paper), may become more relevant. One of our most important aims is to avoid the eradication of many of the beneficial termites. With continuous population growth more land is cleared for building development, which in turn increases the amount of land used for crop production and grazing as well as using trees for timber and fuel. The effects of deforestation have long-reaching effects on biodiversity. Termites help to increase mineral recycling as well as increasing the porosity of soil to water and air. The loss of different termites, as well as other insects, due to deforestation or from mortality caused by the use of insecticides to eliminate crop pests in cleared areas may not seem to have any marked effect initially. However, when taken as a whole with other similar situations worldwide, this could have important repercussions in the future where soil fertility and structure become affected.

Appendix 1

Collection and Identification

Identification is important for deciding on taxonomic relationships between different termite genera and species. It is also important to determine the species of termite when examining termite behaviour, and even more essential in planning any control methods that may be needed.

Termites are most easily found in the cool early morning and late afternoon. Cloudy weather or after rain is also a good time for collecting. It is important to make sure that records are accurate whilst collecting. This is especially important in the case of damaged plants where the termites present may not be the actual ones that have killed the plant but another species that has come along to feed on the same plant afterwards.

COLLECTING METHODS

Firstly, look underneath soil runways. Lift up dead material or plants and break open damaged wood or plant material. To prevent termites returning one can cover the entrance hole with a coin. Using forceps collect as many different termite castes as possible. Soldiers are the main caste used for identification, so it is important to collect as many of these as possible. There may be several different sizes of soldier or worker present that should be collected, but some termites may not have soldiers. Collect other insects that may be living with the termites (termitophiles), as these can aid in the identification of some species and are of interest to termitologists. After rains, alates may be collected at lights or from their exit holes (see Appendix 2).

Place the termites into tubes containing 80% ethanol, kerosene or alcoholic spirits. Dried specimens are useless for identification as soft external parts become distorted and brittle, giving inaccurate measurements.

Write the country, region and place name of collection, type of damage caused, full date and the name of collector on a label and place in the tube. If you wish to include more information add a number to the label and make notes in a notebook against this number. If the termites were collected on an unknown plant this can be collected too for identification later. If the termites are fungus-growing, collect the fruiting bodies, or take a photograph of the nest and the fruiting bodies. Modern technology can provide the exact location of collecting localities using global-positioning hand-held receivers linked to satellites. This gives an accurate position, but it is also important to classify the topography of the area and vegetation present so that other comparisons can be made at a later stage.

If no soldiers or alates are found while collecting, and the termites are not a soldierless species, then it may be advantageous to keep small colonies alive, which is possible with many of the lower termites (e.g. Kalotermitidae) so that these castes can develop.

IDENTIFICATION

Several useful keys exist for termite identification (see Taxonomic References). New computerized keys are being developed for termite identification. Larger species of termites can be examined in order to determine the genus by using a ×10 hand lens. Other species will need higher magnification using a binocular microscope. Termites are normally examined in 70–80% ethyl alcohol in small dishes. One method is to use a small plastic petri-dish containing sponge or other material such as Plastizote on which the specimen can be placed. This also allows the specimen to be pinned down for examination.

Fine watch-makers' forceps are useful, as are needles or pins, to move or hold the termites in position. A calibrated micrometer eyepiece can be used for measuring different termite characteristics, e.g. head length, etc. For measurements the part of the termite to be measured must be placed down as flat as possible and, if in alcohol, must be completely under the surface to prevent any reflection of light. An example of standard measurements is shown in Fig. 7. A narrow source of light (e.g. originating from a fibre optic system) directed at the termite is most useful.

Soldiers are important for identification as their mandibles and head shape can be a defining characteristic for different genera and species. Mandibles have to be 'opened out' for close examination and measurement. This requires cutting of the muscles on the inside of the base of the mandibles (Fig. 7). Identification of workers and alates is difficult and often only possible to the generic level. The wings of alates can be used for identification to generic level, but need to be mounted flat in Berlese on a slide with no folding or air bubbles present in order to see them properly. The arrangement, shape and number of teeth are especially important for these identifications (Ahmad, 1950).

Also useful for worker identification, especially where there is no soldier caste present for a species, are gut characteristics (Fig. 15). Different positions, sizes and modifications of the gut are important, as are the arrangement of chitinous teeth that break up food inside the gut. The arrangement of teeth inside the enteric valve is especially useful, but they need to be mounted in Swanns Berlese on a microscope slide for examination. Where termite specimens are found to have dried out they can be reconstituted in order to help with identification. The termites are warmed in a solution of *c.* 2% potassium dihydrogen orthophosphate and left for at least a day.

Where a piece of wood or plant material has been collected and no living termites are present, it may still be possible to identify the genus from the old head capsules that remain. These, along with the mandibles, are usually the only parts that have not decayed or been eaten by other organisms such as mites and remain after death. The form of faecal pellets and the kind of damage may also help to indicate which genus was originally present. New biological techniques, such as studies on cuticular hydrocarbons (Haverty *et al.*, 1991), genetic comparisons (Korman and Pashley, 1991) and DNA hybridization studies (Broughton and Kistner, 1991) will in future help with the taxonomy of some termites that at present are morphologically indistiguishable and help to confirm or question present classification.

Storage of Collections

Collections of termites are normally stored in small stoppered glass tubes containing alcohol. Rubber, plastic or cotton wool can also be used as a stopper. If the stopper is made of rubber then it should be alcohol resistant. Plastic stoppers can have a small hole pierced in the top so that air can be removed when they are full of alcohol. These tubes are placed in sealed labelled jars also containing alcohol (70–80%) plus a small amount of glycerine to prevent drying out. A thin layer of cotton wool can be placed in the bottom of the jar to prevent tubes breaking if dropped inside. Glycerine can also be added to the top of the jar to ensure an air tight-fit when the lid is replaced. Labels can be placed inside or outside the jar to indicate the country of origin and/or the name of the specimens inside. Writing on the labels must be in permanent ink or pencil.

A constant check must be made of the level of alcohol inside the jar and also inside the tubes. A record book should be kept containing details of the contents of each tube, the origin and date of receipt. Each record should also be given a specific reference code.

Sending Specimens away for Identification

Collections should be split into two before sending off for identification if enough termites have been collected. One half of the collection can then be donated to the identifier for his own collection. This also saves packaging and return postage, which will help in obtaining a quicker response, especially if there are many termites to return. Cotton wool can be placed inside the tube if it is not full of termites to prevent damage during transit. Stoppers to tubes can be sealed with alcohol-resistant tape or low-melting-point wax. Some form of packaging is needed around the tubes (e.g. bubblewrap) which should then be placed inside a plastic bag to prevent possible leakage. Also some packaging (polystyrene chips/shredded paper, etc.) should be placed inside the container or box. The word 'Fragile' should be written on the outside of the box, and the sender's address and any customs requirements stating 'dead insects of no commercial value' should be added. Several organizations exist that can deal with termite identifications. The four major ones include:

International Institute of Entomology, 56 Queens Gate, London, SW7 3JR, UK (an institute of CAB INTERNATIONAL)
National Museums of Kenya, Department of Entomology, PO Box 40659, Nairobi, Kenya.
American Museum of Natural History, Central Park West at 79th Street, New York, NY 10024, USA.
CSIRO, Division of Entomology, GPO Box 1700, Canberra ACT 2601, Australia.

For invertebrate termitophiles (those living with the termites in their nest), contact addresses are:

Dr Henry Disney, University Museum of Zoology, Downing Street, Cambridge CB2 3EJ, UK.
Dr David Kistner, Department of Biological Sciences, California State University, Chico, CA 95929-0515, USA.

Many other organizations in different countries have specialists on termites for different regions. A complete list of termite specialists or other information can be obtained from:

Prof. N.-Y. Su, International Isoptera Society, Ft Lauderdale REC, University of Florida 3205, College Avenue, Ft Lauderdale, FL 33314, USA.

Appendix 2

Culture Methods

Becker (1969) gives good examples of termite culture methods. The easiest termite to culture is *Kalotermes* (family Kalotermitidae) since it is more resistant than tropical termites to drought, low temperatures, food shortages and toxic chemicals. It is therefore a good indicator termite of toxicity and resistance. *Neotermes* of the same family and *Zootermopsis* (family Termopsidae) can also be used. Of the family Rhinotermitidae, *Coptotermes, Reticulitermes* and *Psammotermes* are the most useful termites for tests, for drywood termites (Kalotermitidae) *Cryptotermes* and *Incisitermes* can be used. All these termites can produce supplementary reproductives. Some species of *Nasutitermes, Coptotermes* and *Microcerotermes* in some circumstances, depending on environmental conditions, size of colony and caste ratios, may also produce supplementaries. Other termites kept in culture not producing supplementaries include *Heterotermes, Anacanthotermes, Cubitermes, Macrotermes, Microtermes, Odontotermes* and *Trinervitermes.* These may have a royal pair or have a limited lifespan with no form of replacement. For testing it is better to use several genera as their tolerance may vary.

CULTURE ROOMS

For tropical termites culture rooms should normally be at a temperature of 28–32°C (29°C optimum) with a humidity of at least 80% RH. Other temperate species such as *Reticulitermes* have been successfully kept at 25–27°C in the laboratory. Very low or high temperatures can affect feeding and reproductive rates. *Psammotermes* can tolerate up to 35°C, while *Zootermopsis* prefers 18–20°C and can therefore be kept at room temperature. All termites like a high humidity, which is usually above 85% RH. Termites are especially sensitive to low humidity due to the lack of sclerotization, especially in workers. Humidifiers or open water tanks can be used in rooms and small jars of water can be placed inside containers that hold cultures in jars.

CONTAINERS

Jars and Tanks

Figure 57 shows a range of containers that can be used to keep termites. Containers should be small enough for easy handling but should also have a wide mouth in order to be able to reach inside. In small containers and sandwich boxes it is better not to exceed 500 termites. Large containers should not be used for small colonies as they may not be able to cope with any fungal growth. As the colony grows, the termites can be transferred into larger containers or tanks. It is easier if very large tanks are placed on a trolley or board with wheels so that they can be moved around. A sheet of glass or perspex can be placed on the top to prevent termites escaping.

Plate Nests

Using plate nests, termites can be counted and their behaviour observed under different conditions (Pearce *et al.*, 1991). To make a plate nest, two identical glass sheets are needed. The lower sheet is separated from the upper one by glass strips along three sides (Fig. 58), which are stuck down along the edge of one of the pieces using a strong glue, such as Araldite. It is important that the thickness of the glass strips should only slightly exceed the height of the insects so that they can be seen.

The upper glass sheet can be divided into several pieces, which allows easy access to one area of the nest without disturbing the others. If a small glass sheet is required, microscope slides can be used as the upper pieces. Wood or soil can be placed inside the plates and the open end sealed with cotton wool or gauze.

SETTING UP COLONIES

Wood as Food

Softwoods that are susceptible to colonization by termites, such as pine or spruce can be used in setting up colonies. Hardwoods are preferable for *Nasutitermes* spp. If dead or rotting wood is collected from the field then it needs to be dried overnight in an oven to kill microorganisms and other predatory insects or parasites living inside.

Fig. 58. Large glass plate nest and supports used to maintain colonies of *Microtermes*.

Wood used for drywood termite culture should be conditioned for at least 48 h in a high humidity at 28–30°C before use. This can be done in a sealed chamber containing jars of water. Wood for other termites can be placed in water for half an hour and then excess water shaken off it.

When placing wood in containers, a piece of the original wood should be included with the new culture wood so that the termites can adapt to the new food source. It is very important when using containers such as glass jars to place some very small pieces of wood in the base before adding the larger pieces. This will allow termites that fall to the bottom to climb back onto the larger pieces of wood. If this is not done any termites trapped at the base would eventually die and cause fungal or mite problems.

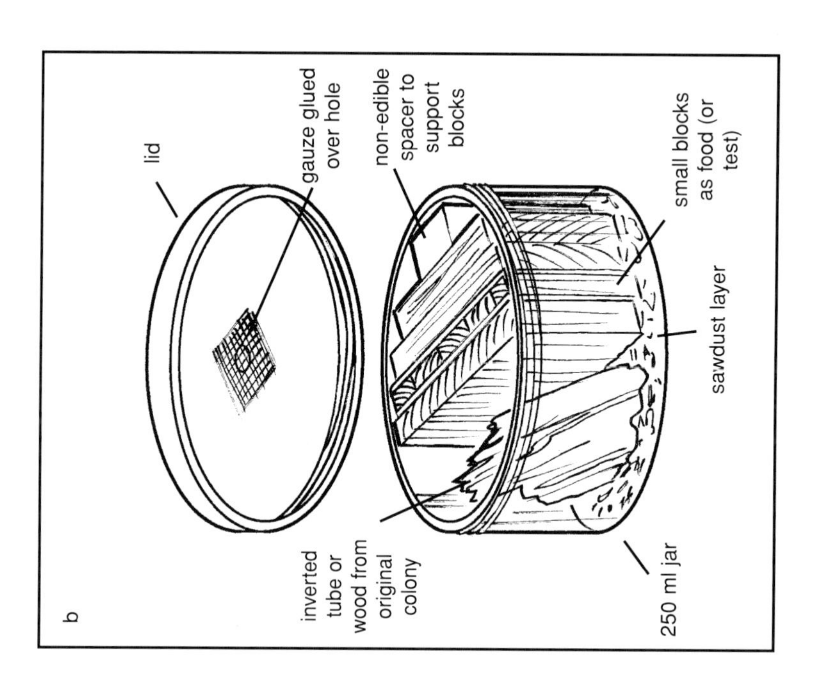
b
lid
gauze glued over hole
non-edible spacer to support blocks
small blocks as food (or test)
sawdust layer
250 ml jar
inverted tube or wood from original colony

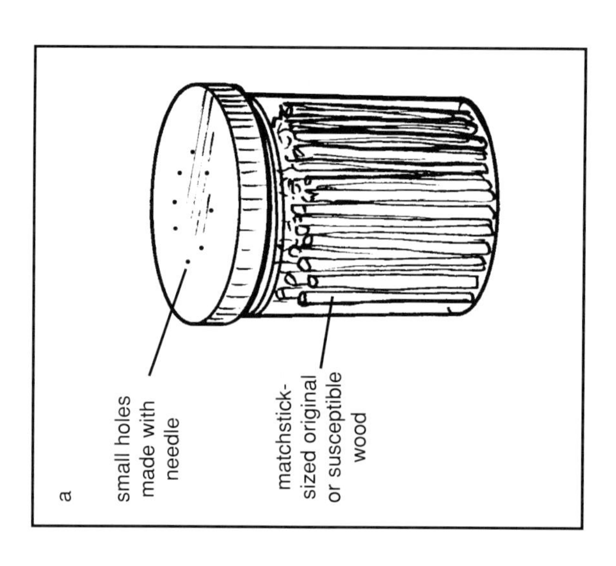
a
small holes made with needle
matchstick-sized original or susceptible wood

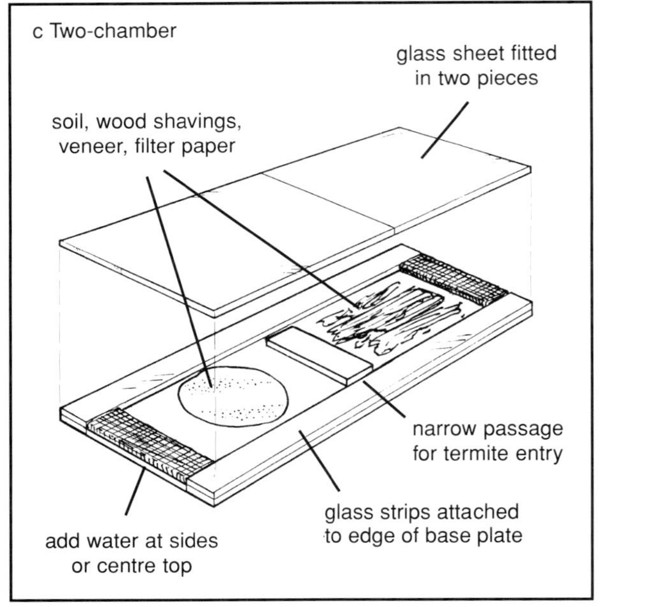

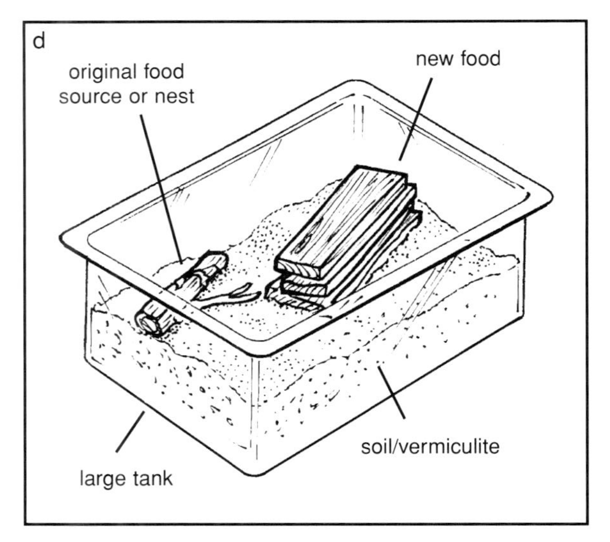

Fig. 57. Culture of termites: (a) small, glass-stoppered vial with wooden matchsticks as food source; (b) wood-dwelling termites; (c) plate nest for wood- or soil-dwelling termites; (d) for soil-dwelling termites.

Matrix for Colonies

For subterranean termites, vermiculite can be used instead of soil in containers. This can also be mixed with sterile soil, sand, nest material or even a food material such as sawdust. Heat-sterilized soil mixed with nest debris has been used successfully for *Microcerotermes*. For *Coptotermes*, vermiculite alone or mixed with soil or nest material can be used. For *Nasutitermes exitiosus*, pieces of crushed nest material should be included. Soil with a high proportion of clay is normally best for those termites that form soil runways, especially the fungus-growing termites. Clay provides more support for tunnels and chambers and material for mound building for nests in tanks. The ideal matrix moisture content is attained when the surface remains damp but not flooded. Wooden slats held together in bundles of six with rubber bands or string can be placed on the soil surface to act as a food source.

Collecting Termites

Collected termites will only be useful for culture if they have the ability to produce replacement reproductives, or if the king and queen are collected as well. Where the nest is in the soil, the royal chamber and contents must be collected. Baits consisting of bundles of wood or other forms of cellulose placed on or partially buried under the soil surface near or under nests can be used to collect termites. Corrugated cardboard placed in plastic tubes, wrapped around susceptible wood or simply rolled up and covered with a plastic bag can prove effective. Glass plates containing cardboard placed on the ground and covered with soil have also proved useful for collecting termites, with little loss or injury (Pearce, 1990). Jars that have been filled with pieces of wood so that they protrude out of the top can be used to collect termites for culture. The jar is placed on the soil surface with the protruding wood under the soil. Termites enter the jar, the protruding wood can then be sawn off and a lid placed on the jar. This is only useful for those termites that produce supplementary reproductives so that a new colony can be produced. Termites living inside logs or timber can be placed, and kept temporarily, in a plastic bag in the dark. Large plastic boxes or bins can be used for nests. When extracting termites from wood, large pieces of damp filter paper can be used to attract and collect termites away from the debris.

With some of the lower termites it may be possible to pool together individuals from different colonies of the same species to form new viable colonies. It is best to test whether this is possible first with a few termites and then observe their behaviour for at least a week before combining large numbers. Chilling (i.e. cooling) termites in a fridge may sometimes help to reduce any aggressive behaviour and improve mixing. Pooling termites is not a good idea if termites are soon to be used for testing or bioassays.

Placing Termites in Containers

Termites should be introduced to containers slowly. If many soldiers appear to be present when extracting, it is better to remove some of these to reduce the amount of time that the workers need to spend on feeding them rather than on colony formation.

Nest or Mound Collection

The royal chamber and many workers need to be collected from dispersed nests and mounds. Whole mounds can be collected into bins or tanks (Fig. 59). It is also important to collect workers not found in the central part of the nest as these are the foragers and builders. If the nest contains fungus-combs, these should be collected and initially included with the new colony. It is important to remove comb that has become unattractive to termites and to replace it with fresh comb.

Fig. 59. Laboratory colonies of *Macrotermes*. Large and small tanks with feeding chambers connected by tubing.

Sub-culture

Colonies of some species of the wood-dwelling termites, such as *Cryptotermes*, can be produced from only a few workers. However, it is often better to use more workers to increase social contact and trophylactic exchange to avoid water loss and build up the colony numbers faster. For termites of the family Kalotermitidae, more than a hundred individuals can be used, for Rhinotermitidae more than two hundred and for others more than a thousand. Where alates are present, numbers can become seriously depleted, so it is best not to sample at this time.

Colonies from Primary Reproductives (Alates)

If alates are seen flying during the day or night there is a good chance that they will fly again at the same time on subsequent days or nights. Alates collected at a light source or whilst flying from colonies must be sexed, paired and placed in soil or wood as soon as possible.

Alates can be picked up by their wings or can be collected more easily by allowing them to run on to a small piece of paper and then tapping them into a funnel placed over a tube. Most alates form tandems after flight, and these can be collected together and placed straight into a tube with a hole in the lid. If the termites are emerging from a flight hole then a plastic bag can be placed over the top to collect them.

The male differs from the female in having of a pair of sub-anal styles. The female, unlike the male, has some of the abdominal segments fused (Fig. 4). Wings can be removed by hand if termites have not already lost them. The wings are raised over the head using forceps (avoid pulling the wings sideways, which could injure the termite); this causes them to break off at the suture.

Setting up Colonies for Wood-dwelling Termites

The production of a large-sized colony from a pair of alates in the lower termites can take several years, so many small colonies need to be set up to obtain enough insects for testing. Alates can be placed in small tubes, in holes drilled in small blocks of wood or in slits in wood veneer placed in a glass plate nest (Steward, 1983).

Setting up Colonies for Soil-dwelling Termites

Sterile soil is placed in a tube or small petri dish (45 mm diameter) and dampened but not soaked. A small hole is made in the soil which will allow alate pairs to tunnel down and to lay eggs and produce young. These dishes or tubes should be placed in a container that contains a humidity source, such as a jar of water, and kept in the dark at the appropriate temperature (e.g. in an incubator). No food is needed at this stage as the parents provide food from their body reserves, but some distilled water should be added to keep the soil moist (but not soaked). Once the first young start to forage food, small wood flakes must be provided on top of the soil surface. The alates of some fungus-growing termites (Macrotermitinae), such as some species of *Macrotermes* and *Microtermes,* may carry an inoculum of the fungus species *Termitomyces* with them when they leave the nest. This is needed to break down food collected by the first foragers so that they can survive. Other members of the Macrotermitinae normally search for spores at the soil surface. This means that for these species original fungus-comb needs to be added to the colonies.

Mortality is often high for alate pairs and nymphs in soil. It is therefore better to set up as many colonies as possible to allow for at least a 50% mortality rate. Once the first workers are produced then mortality drops rapidly. Useful references on culture are Watanabe and Noda (1991) and El Bakri *et al.* (1989a).

Water Requirements

Daily checking is needed for new colonies, especially in small containers where the soil can easily become dry. Weekly observation is needed for large colonies to make sure the soil is damp. Water can be supplied directly to the soil or by the use of open jars of water in closed containers for small cultures. Khan (1980) summarizes the effect of humidity on the activity and development of some of the more common termites used in the laboratory.

Problems

Various problems can occur with termite colonies. In all cases, unhealthy cultures should be isolated or sealed away from other colonies in the culture room to prevent the spread of infection. Also, where an infection has killed off a large number of termites it is often worthwhile to set up a few small colonies from this original colony. This may ensure that the whole colony is not lost. Most chemicals that could deal with the problem are also toxic to termites, so cannot be used.

Fungi can attack wood in humid conditions. Termites can deal with this in small amounts, and for some termites this is beneficial. However, some rots are toxic and if excessive fungal decay occurs, new wood must be added to the colony or, in serious cases, the termites need to be transferred to a fresh food source.

Mites, especially in their resting stage, cause death of termites by attaching themselves to the legs or mouthparts. The termite then has restricted movement and may be unable to groom or feed. The presence of mites and some dead termites in a colony indicates poor colony health. Healthy termites should be selected and transferred to a temporary new container with fresh food and soil, if subterranean. They can then be transferred to a more permanent container when mite numbers have decreased.

Cannibalism is common in small colonies and where the food supply is low. If the numbers have become small and the workers can form supplementaries it may be worthwhile isolating a few individual workers to start off a new colony.

More details on culture methods can be found in *Laboratory Culture and Experimental Techniques* (M.J. Pearce, 1977, Natural Resources Institute, University of Greenwich Publication).

Appendix 3

Monitoring Methods

SAMPLING

There is a wide range of sampling methods that can be used for collecting termites. Extraction from soil can be carried out using Tulgren or Berlese funnels, but a more effective method is by hand-sorting or dropping soil into water so that the floating termites can be collected.

Several different kinds of baits can be used for collecting or monitoring termite activity/distribution. Litter bags can be used on the ground, but the mesh size must be specific for termites and not predators. Also, the mesh must not be made of a material that can be eaten by termites. Wooden blocks or boards can be used (see Monitoring Foraging, p. 139), as can corrugated cardboard or other waste material attractive to termites. Wood or paper placed in containers can form a means of trapping termites so that when the container is lifted the termites remain inside.

Other methods of monitoring distribution include pitfall traps and examination of vegetation within 50 m^2 quadrats. The termites present in timber of a particular size can also indicate termite abundance and marking with dyes, fluorescent compounds, spray paint or radiolabels can indicate population and territory size.

Populations of mounds can be found by fumigating and sub-sampling, the termites being extracted by sieving, hand sorting or flotation. Collecting alates (winged termites) may indicate the presence of termites in an area which had not been detected by the other methods. Light and sticky traps can be used to trap the alates. The population of alates in a mound can be determined by fumigation using methyl bromide.

MONITORING PLANT MORTALITY

Examples are given here of how to monitor plant mortality in the field. This is useful to gain an indication of the degree of termite attack and its variability throughout the season. Plant mortality can then be related to environmental, and other, factors present. Monitoring is also important for assessing the usefulness of different control methods. However, it must be noted that damage assessment of plant mortality may not necessarily be directly related to yield (Wood and Cowie, 1988). Some plants may benefit from the death of a neighbouring plant and increase their growth and, therefore, yield.

Baiting in the field also provides an indication of termite activity, which again is important when trying to relate the effectiveness of different control methods being tested.

Before carrying out trials and setting up monitoring procedures, it is essential that different aspects of the whole control programme are reviewed. The results of any findings will then have significance to all concerned, and not end up as a report where recommendations are not practical due to costs or labour/equipment involved. Some of these considerations are listed in this section.

The following methods for monitoring plant mortality outline those used by Wood *et al.* (1987) and Tiben *et al.* (1990).

A number of sample rows with a good number of plants should be chosen, and at least one plant left at either end as a discard. The number (initial plant count) should be recorded and the rows labelled. Mortality can be recorded on sample rows with the end plants on each row being discards. Records should be made every two weeks throughout the season. Plants are examined for wilting and the area checked for runways (if several species of termite are causing the damage try to find out which species is attacking the plant, e.g. by looking under the runway). Where the plant is beyond recovery it is removed from the ground and checked for damage and for the species of termite responsible. Termites, if present, are collected and labelled noting their location. (Note that before starting the experiment some knowledge of the termite species present in the area should have been gained.) It is important to make sure that the death of the plant was due to termites and not caused by another pest or disease. Crickets, bacterial rots and virus diseases can be alternative causes of plant death.

In the Sudan trials on cotton (Tiben *et al.*, 1990), 16 rows were sown per plot with 15 stands per row. Of these, 6 rows with good germination were chosen for monitoring, with 14 stands in each row. At the end of the experiment all plants were pulled out of the ground and examined for root damage. If the crop produces pods or tubers underground, some estimate of damage will be needed on removal. Dead groundnut plants should be examined for root attack and to see whether they are penetrated or scarified. The total number of pods should also be recorded.

Percentage plant mortality is based on initial plant number. Mortality results can be added to give a cumulative plant mortality for the season. This can be plotted against sampling dates as for the wooden bait results (see Monitoring Foraging p. 139). Means derived from mortality results can be tested for significant differences and arranged in order of significance using multiple range tests or other statistical methods. Results can be represented under the headings:

Treatment % plants killed Yield (kg ha^{-1}) Value of crop Grade

The value of the crop can be given for different grades (quality) for comparison. Many different expenses have to be considered when costing. Examples for cotton include:

- Cost of seed
- Cost of empty sack
- Cost of seed dressing, or other insecticide, and for treatment
- Cost of labour, maintenance and fuel
- Thinning labour costs
- Weeding costs

Benefit-to-cost ratios are used to indicate what is gained from the trial and whether the method is useful. An example is given in Table 5 showing aldrin soil application compared with seed dressing for cotton.

Table 5. Benefit-to-cost ratio.

Treatment	Season	Cost	Return	Gross margin	Benefit-to-cost ratio
Aldrin soil application	1982/83	10.0	60	40	6.0
Aldrin seed dressing	1983/84	1.0	60	59	60.0

In this case seed dressing has ten times the benefit-to-cost ratio of soil treatment. It also was safer, so is readily recommended.
Cost = insecticide, labour, equipment etc.;
Return= money/hectare based on how much money gained per kilogram cotton;
Gross margin = return − cost;
Benefit-to-cost ratio = returns divided by cost.

MONITORING FORAGING

Farmers and pest controllers need to know which termites are present and where these are abundant, e.g. by collecting, observation and baiting at different times. One successful method is the use of pine softwood baits (10 cm × 2 cm × 2 cm) for monitoring termite foraging. Any susceptible material, even bamboo and cardboard discs, can be used. As well as tightly coiled rolls of single-faced fibre board and tubes from toilet rolls.

Before setting up trials it is useful to carry out initial baiting in the trial area, especially where termites do not appear to be spread over the whole area. This will also give some idea of the termite species present and the rate of bait removal so that exposure times can be adjusted if needed. Once an area is chosen, baits can be put down immediately after planting. For wooden baits, a shallow groove is first made in the soil and the bait place into it. This gives it less chance of being displaced. Baits are often accidentally removed, moved by animals or can even be removed for fuel, so it is important to have a person looking after the area. Baits should be placed between the plant rows with at least one bait placed every 2 m in each row. Baits should be placed between rows with at least one sample row on one side (for example a 12 × 12 m plot with six sample rows with five baits in each row, giving 30 baits per plot). A reasonable time interval to examine the baits is every two weeks, but this can be modified after observation. The baits should be replaced by fresh ones, as old baits could be rejected.

Foraging is estimated by the presence or absence of termites. The bait may have been eaten (attacked) but there is an absence of termites. In this case, the bait is recorded as attacked, but still represents activity during the two-week period. Results of bait attack can be represented as follows:

Table 6. Bait attack.

	Number of baits			
Plot row	Attacked by termites	Attacked but gone	Not attacked	Disturbed or missing
1.				
2.				
3.				
4.				
5.				
6.				
Total				

The percentage attack found over the two weeks is given by the total number of baits attacked, or by the number of termites present, divided by the total number of baits used minus those missing. If several different species of termites are found on the same bait then these must be recorded on the table, e.g. 1/O 2/M (i.e. one bait with the termites *Odontotermes* present and two with *Microtermes* present).

A graph of the cumulative percentage of attack over the season can be drawn from the results (Wood *et al.*, 1987; Tiben *et al.*, 1990). One can also statistically compare bait attack for different treatments at the same sampling date and seasonal attack on the baits used as controls.

Survey of an Area for Termites

As well as baiting, examining nests and wood/litter on the ground and dead wood branches of 2 cm diameter or larger, on individual trees, can be cut down and examined for termites.

Testing Chemicals

When evaluating chemicals in wood for possible use as barriers, treated wooden baits can be placed in randomized blocks at several different field sites. As well as using small wooden baits for monitoring termite activity for testing the effectiveness of different chemical soil treatments, wooden ground boards can be placed on the soil surface. Chemicals can also be applied around the perimeter of concrete slabs, which are then lifted to see if termites are present. This test is important as a screening test for effective control methods.

IMPORTANT CONSIDERATIONS FOR SUCCESSFUL FIELD TRIALS

1. Before starting it is important to find out basic information about the pest and its natural enemies (from literature and from talking to other scientists and farmers).
2. One must be aware of any regulations and policies of the corresponding government and other organizations that may have an effect on carrying out trials. Once these have been discerned, arrangements can be made for long term cooperation.
3. An assessment must be made of the effectiveness of the proposed control methods. This includes checking of control costs against economically acceptable levels of damage and loss. This should be assessed throughout the project. Also, the correct code of conduct needs to be applied for the use of the control method, with training in safety for those involved.
4. A viable site should be chosen where labour is no problem and transport is easy.
5. The methods used for monitoring need to be decided and training given to extension workers and farmers.
6. Organized feedback of results needs to be developed throughout the project so that new proposals and their funding can be agreed.

THE VALUE OF MARK–RECAPTURE METHODS

A single mark–recapture experiment can be used and an estimate on the population made using the Lincoln Index, which is:

(Basically), the estimate of population =

$$\frac{\text{Numbers captured, marked and released} \times \text{total numbers captured (2}^{\text{nd}}\text{ time)}}{\text{Number of marked individuals recaptured}}$$

An example of the use of this method is seen in Jones (1991). There is, however, a large standard error using this method so a triple mark–recapture method procedure using a weighted mean model has been used to try and improve the accuracy of foraging population estimates (Su, 1994b).

Thorne *et al.* (1996) have evaluated the use of mark–recapture methods for field colonies of *Reticulitermes* and have concluded that one should interpret foraging population estimates with great caution. Many of the methods assume random foraging, which has never been demonstrated and is certainly not true if there is a preferred food source. Dyes can be lost naturally or by other causes, e.g. poor health and moulting. Recapture may be affected by size and age. The application of the use of dye methods to evaluate changes in population over time is also questionable. A great deal more information is needed on the behaviour and colony organization of each species of termite before definite population size estimates can be achieved.

Appendix 4

Laboratory Tests Using Termites

Termites can be collected from a piece of wood by tapping the wood so that they fall into an underlying dish. Damp filter paper can be placed in the dish to attract termites on to its surface. These can be tapped off into a new container or used for testing. Dead or injured termites are removed with a moistened, fine paintbrush. Testing time depends on the sample size, number of insects and kind of sample. For toxicity testing, termites need to survive in experiments for at least 14 days. To increase survival for the fungus-growing termites, small pieces of fungus-comb or 5% sucrose can be added to the dish. Counting termites for experiments may be more accurate than weighing, especially where different size castes are present.

CONTACT POISONS

Dusts

A thin layer of dust (of known particle size) is sprinkled on to the base of a 95 mm diameter petri dish and 50 termites are allowed to walk over it for 1 min. The dish is then shaken to coat the termites with dust and they are then poured out on to a filter paper allowing them to remove any excess dust. Termites are then placed in a dish with a food source, i.e. filter paper, and mortality is monitored. This method can also be used to examine transfer to another group of termites using different proportions of termites or exposure times to the dust.

Insecticide dusts or liquid can be added to soil or sand (of a known uniform particle size) which has been dried to constant weight. Treated soil (1 g) is placed in each half of a petri dish with a groove left between. Termites are placed in the centre of the groove and the number of termites found on each side is recorded at 15 min intervals. Repellency is considered present when at least three-quarters of the termites are on the untreated side on several replications. Bacto agar (1.5%) mix can be added to the soil to stabilize it if required.

Several other methods of testing treated soils exist. All involve placing the test soil between a container containing the termites and one containing a food source. The soil may be held in a tube (Tsunoda, 1991) or a small container (Grace, 1991). Su *et al.* (1993) and Grace *et al.* (1993) have used agar to hold treated soil in place in the centre of a tube. Again, the termites are placed on one side of the agar and on the other, the food source, so that the termites are attracted through the agar into the soil and out of the other side to the food.

Topical Application

To test the effect of a contact insecticide, chemical is added to the upper thorax of termites under a microscope using a microapplicator (0.5 μl of solvent, e.g. acetone, containing insecticide). The termites can be weighed to calculate the required dose ($\mu g\ g^{-1}$). Treated termites are then placed with a fresh food source and survival recorded.

FEEDING EXPERIMENTS (TREATED PAPER OR SAWDUST)

Tolerance

Tolerance tests determine the concentration of chemical or pathogen that will not kill the termites immediately, but will be transferred to nest mates causing reduction in fitness and eventually death. An ideal concentration is one that causes less than 10% mortality for the first 3 days and 100% mortality by 14 days.

Termites can be kept on filter papers (42.5 mm diameter, Whatman No. 1) or sawdust can be used instead (>20 mesh softwood) mixed with a known concentration of test solution.

For filter papers 0.15 ml of the test solution is used, with distilled water used for controls. Ten termites are then placed on the treated paper inside the dish, a filter paper is glued to the inside of the lid, moistened and the lid then replaced. Mortality is recorded and any dead are removed. Treated papers can also be used in small colonies in plate nests or placed in dishes connected by tubing to larger colonies to see the effect on colonies.

Timed Exposure

The same method as for the tolerance experiment is used, but the termites are removed to untreated paper after different exposure times.

Transfer Test

The same number of termites are used in each dish, but with various combinations of termites that have been treated differently. Some termites may have been treated with a chemical, and others not treated, just dyed with a marker. The transfer of a toxic chemical from treated to untreated (dyed) termites can then be monitored.

Preference Test (Consumption)

In a no-choice situation the test sample may be eaten by termites, but given a choice, this may not be the case. To test this theory 20 mm diameter filter papers (Whatman No. 1) can be treated with 0.03 ml of test solution to give the concentration range that excludes those concentrations causing high initial mortality and mortality equivalent to the control. Papers are treated with distilled water as a control.

Papers can be placed on top of the soil in the container or placed inside petri dishes which are connected to a colony by tubing. The amount of feeding or spotting with faecal material can give some indication of the preferences.

Preference Test (Visual or Video Recording)

Initial observations are important especially where one chemical may be repellent. Single petri dishes (45 mm diam.) can be used or large 95 mm diam. plastic petri dishes divided into three compartments, to give three replicates per dish. Filter papers (20 mm diam.) are treated with 0.03 ml of test solution. One control and one treated paper are placed in a petri dish with a gap between them. Alternatively, a single paper can be cut in half and one side used as control and the other treated.

Termites can be counted at regular time intervals by observing them through a piece of glass covered in a red plastic film, which prevents the disturbance that would occur in bright light. The number of termites on each of the papers is recorded after 30 mins and then at hourly intervals. To monitor over 24 h, a time-lapse video recorder can be used with an infra-red camera and a lamp with a red filter for illumination.

FEEDING EXPERIMENTS (WOOD)

Wood blocks of not less than 1 cm cubed should be used in feeding experiments so that the termites can find them and so that there is less chance of fungal infection on all sides. Samples should be taken at random from more than one tree. Assessment includes the amount of tunnelling present, the wood on which the termites are found and mortality.

Several different ways exist of presenting and testing wood samples. Tests should run for a minimum of six weeks and exposure time depends on the rate of damage to the specimen compared with controls. Blocks can be removed weekly for assessment.

The simplest approach is to place a small weighed block on filter paper in the centre of a petri dish. If a matrix is used that contains termites, the block can be placed on top or buried under the surface. If treatments or wood constituents are able to leach into the matrix, four pins, or a glass/plastic ring can be used to raise the block above the surface. To see if any damage has occurred the blocks can be placed against one side of the container. Veneer can be placed inside the plate in plate nests with a central slit cut out to accommodate the termites.

Surface Ring Method

A glass or resistant plastic ring is placed on top of the test sample (e.g. wood) and termites are placed inside (Becker, 1969). The ring can be placed so that it nearly extends over the edge of the sample, the termites then have to attack and eat the edges away to tunnel through.

If the species of termite does not survive well away from the parent colony, a length of tubing can be attached to the container that holds the colony. The other end is fixed to an inverted petri dish which is placed on top of the test sample.

Another method that is used involves placing a layer of plaster of Paris at the base of a glass or plastic ring. The ring and plaster is then placed on to damp cotton wool so that the plaster can absorb moisture and provide a moisture source for the termites (Tsunoda and Nishimoto, 1986; Tsunoda, 1991). This produces an arena for adding test woods and termites.

Glass Tube Method

Another method is to set up a small colony inside a glass tube open at either end, the end of which has been stuck with Araldite to the upper surface of the test sample (British Standards Institute, 1989). A small piece of wood shaving or cardboard disc is dropped to the bottom of the tube to attract the termites downwards.

Vermiculite or soil is added to the tube so that it is one-third full and a piece of original food source is added before the rest of the vermiculite is added. Water is added to dampen the vermiculite and at least 200 workers, a few soldiers and a few other castes added. Metal foil or gauze can be used to cover the top of the tube. After about two months, or sooner if the termites have tunnelled through the test source, the test material is examined and tunnelling recorded and the number of termites left counted.

Plate Nests

Single pieces of wood veneer can be tested using plate nests. Survival can be monitored daily using this method and it can also be used to examine colony foundation.

Wood on Ring Method

Test wood is placed at the bottom of the test container under the matrix. A plastic glass ring is placed at one side of the jar and the test block placed on top of the ring so that it rests against the side of the container (British Standards Institute, 1990).

After adding termites, their position and activity in the jar is observed. At the end of the experiment the attack on wood can be ranked by: 0, no attack; 1, attempted; 2, slight; 3, average; and 4, strong (British Standards Institute, 1989, 1990).

Multiple Sample Method

Sandwiches of veneer held together by clips or adhesive tape can be used.

For larger tests flat pieces of wood can be placed next to each other in jars or tied up in bundles and placed on top or buried under the matrix in tanks. Small blocks of different woods can be tested as pairs or arranged radially in a petri dish. Termites are placed in the centre and the number of termites recorded on the blocks at 30 min intervals. For the latter case, the most attractive block is removed first and then subsequent attractive ones, so that a ranking of woods can be obtained. For termites which cannot survive long away from the mother colony, tubing connected from the colony through the base of the petri dish to the centre of the test arena can be used.

A method using a plastic ring on a plaster of Paris base (Tsunoda, 1991) can also be used for testing treated and non-treated wood blocks.

TESTING OTHER MATERIALS

Paint

Three coats of paint can be applied to either side of a filter paper (2 cm diam.). This is allowed to dry for several months to remove volatile effects caused by the solvents in the paint, before being placed in a petri dish and tested with termites.

Cables

Cable or wire is sealed at both ends with a metal cap or foil and placed in a large colony of termites under a food source, e.g. a susceptible wood.

For small-scale testing a chamber can be constructed across the piping with the food source above the pipe and termites below. Termites have to eat past the edge of the cable to reach the food source. Assessment is by volume loss and the degree of internal cable exposure.

Plastic Film or Dam Lining Testing

These can be wrapped around a block of susceptible timber or rolled into a hollow tube *c.* 2cm diam. They can then be tested as for cables.

Material

This can be sandwiched between foil or plastic, where one side of the sandwich contains holes so that termites can spot-feed when kept on top using a glass ring.

Preservatives

Weighed paper, wood wafers or blocks of susceptible wood can be dipped in preservative or it can be applied by brushing. Small blocks can be split in half to allow access to the interior of the block. The cut surface is placed downwards on the substrate. Reweighing gives the weight of preservative applied. Hardness can be tested or the effect of various numbers of coats of paint. Untreated wood and wood treated with solvent should also be compared in the test. The effect of different pH on feeding can also be examined using this method.

Diets

Where the diet is liquid, this can be tested as for the methods using filter paper in petri dishes. For a more solid diet, the food source can be moulded or cut into shapes, weighed and used in a similar way to wood blocks for testing. Diet can also be placed in plastic tubes and holes drilled into the base to allow termite access. Cellulose powder with various added constituents can be tested. Dry powder can be mixed with twice its weight of distilled water and other additions are calculated as a percentage of the total weight of the diet.

BIOCONTROL AGENTS

Possible pathogens as an alternative to chemical control have been investigated (Suzuki, 1991). A useful protocol for testing the effects of microbial pest control agents has been published by Grace (1994).

Fungi

Various strains of *Metarhizium anisoplea* and *Beauveria bassiana* can be grown on agar, e.g. Molishes. Spores are released from the media using water and teepol. The suspension is then sieved through muslin cloth. The concentration of spore solution is found by placing a drop under a haemocytometer. An average of ten drops can be used. Termites are placed in a dish and the dish placed on ice to cool the termites. A 1 μl pipette is used to dose the termites on the abdomen with spore solution. Another method is to add a known volume of liquid containing spores to a filter paper in a petri dish and to allow the termites to walk over this. Daily observations are made on the termites and any dead ones removed and placed in separate dishes to see if they have been affected by the fungus.

Bacteria

Bacillus thuringiensis (BT) is a commonly used bacterium in pest control. A suspension of known concentration (viable spores g^{-1}) can be used and is applied in the same way as for fungi. Initial symptoms include sluggishness and failure to respond to touch. Virus application is also similar, here body fluids are examined later for virus particles.

Nematodes

Examples of nematodes that have been tried against termites include *Heterorhabditis bacteriophora, Steinernema carpocapsae* and *Steinernema riobravis.*

Concentrations of nematodes in solution are found using a haemocytometer. Known concentrations are added to damp filter paper and termites are added to see if they become infected. Dead termites are removed to wet filter paper for nematodes to develop. If nematodes are not found when dissecting in Ringer's solution then it is assumed that the termites had not been killed by them.

OTHER USEFUL EXPERIMENTS

Humidity Preference Tests

Gradient apparatus can be used in humidity preference tests. Containers are divided into several equal compartments and a humidity range of 43–95% at 28°C can be obtained using calcium chloride at one end, sodium chloride in the centre and water at the other end (a list of salt solutions that can be used for gradients is given later on in this appendix). In each test termites are placed at one or other end of the apparatus. The passage way is cleaned between experiments using alcohol and records of positions are made every hour. Humidity preferences can also be tested using wood veneers in circular humidity gradients or small plate nests kept in containers at different humidities (Steward, 1982).

Temperature Preference Tests

This experiment can be used to examine different termite preferences between genera, species of the same genus, different castes and termites of the same species taken from different habitats or geographical areas.

An aluminium plate, having one end connected to a heating coil and the other in ice, can be used to produce a temperature gradient. A sealed container can be used to cover the plate and to prevent termite escape. Humidity and airflow can be monitored within the container.

Gravity Testing

Some termites are geonegative (e.g. some *Neotermes* spp.), and others geopositive (e.g. some *Microcerotermes* spp.). It may therefore be important to know the movements of termites for setting up a particular test. One simple method is to use a piece of cardboard or glass covered in graph paper. This is held vertically and termites are placed in the middle. Positions are recorded every 10 minutes.

Chemical Attractants as Trails

A semi-circle bisected centrally by a straight line is drawn on a piece of paper. A sheet of ground glass is placed on top of the paper. Test solutions are applied by micropipette along the position of the circle on the glass and the solvent added along the straight line. Termites are placed in the middle and the area covered with a transparent container so as to prevent air currents.

Interactive Behaviour

Aggressive behaviour shown by one termite to another may indicate that the termites originated from different colonies. To test this, small groups or individual termites can be placed on filter paper in a petri dish and their reactions noted. If the termites normally live inside wood, some fine sawdust can be included and if they live in soil, a small amount of soil can be added. The sawdust or soil may be used by the termites to form barriers so that termites from the different colonies do not come into conflict (Pearce *et al.*, 1990).

Protozoa

These are found in the lower termites (Yamin, 1979a,b). One of the best termites in which to observe protozoa is *Zootermopsis*. The gut can be pulled out from a freshly killed termite using forceps and placed on a microscope slide. The rectal pouch is opened to release the protozoa and a drop of 0.6% saline added to extend their survival. Termites must not be just about to moult or have just moulted as protozoa will be absent.

Examples of some of the common protozoa found in *Zootermopsis* are shown in Fig. 28. The largest protozoa are *Trichonympha* spp. Also present are spindle-shaped *Streblomastix* which are unusual in having a deeply-grooved striated appearance on their surface that is due to long rod-shaped bacteria joined end to end. Smears on a microscope slide can be fixed in Schaudinn's or Carnoy's fluid and stained with Delafield's haematoxylin.

USEFUL PROCEDURES

Calculation of Chemical Dosages for Treating Filter Papers

It is known that a 4.25 cm diameter Whatman No. 1 filter paper will absorb 0.15 ml of liquid. Let us say that the mean weight of the paper (based on 50 papers) is X g. Then for a 10% solution w/w in each filter paper we would need $10/100 \times X = Y$ g. If the product is not 100% pure but only 50% then the amount required is $Y \times 100/50 = Z$. For 100 ml of 10% solution we need $Z \times 100/0.15$. Serial dilutions can then be made by adding 0.1 ml to a 10 ml flask and making up to the mark or 10 ml in a 100 ml flask. Equal volumes are used to halve the dilution.

Preparation of Molishes Agar for Fungal (e.g. *Metarhizium* and *Beauveria*) Cultivation

Molishes agar (for 1 l)

15 g of agar is added to 500 ml water in a litre flask and then heated to boiling until dissolved in a waterbath for 1.5 h. 20 g of sucrose and 0.25 g of K_2HPO_4 is added to 250 ml water in a flask. Then 10 g of peptone and 0.25 g $MgSO_4$ are added to 250 ml water and this is added to the same flask.

The final mixture is then poured into bottles and autoclaved at 15 psi for 15 mins.

Salt Solutions used for Maintaining Different Relative Humidities in Containers at a Temperature of 29°C

Potassium sulphate 97% ± 2%
Barium chloride 89%
Potassium chloride 84%
Sodium chloride 76%
Sodium nitrate 70%
Sodium nitrite 64%
Sodium acetate + sucrose 58.5%
Calcium nitrate 52%

Marking Methods

Dyes, if not water soluble, can be prepared in 1–2% w/v solutions with acetone (controls soaked in acetone). 42.5 mm papers can be left in dye overnight then weighed after acetone has evaporated. The dyed paper is then placed in 50 mm petri dishes with 40 workers. Calco-oil red persisted in termites for 3–4 weeks if fed on dyed paper for 2 weeks. The dye can also be added to sawdust instead of the filter papers.

For *Microtermes* spp. the dye Nile red has proved the most suitable. Aniline blue can be used to trace passage of food from one termite to another. Nile blue and neutral red have been successfully used for marking *Reticulitermes* and *Coptotermes*. Sudan red 7b on Whatman No. 1 paper at 2% wt/wt concentration was found to last for 15–45 days in termites. However it must be remembered not to collect preserved insects in alcohol as dye is lost.

Fluorescent spray paint has been used successfully for marking termites. Red, blue and yellow are readily distinguishable. Dyes for marking *Microtermes* have been examined by Salih and Logan (1990) and for other subterranean termites by Forschler (1994) and Oi and Su (1994).

References

GENERAL REFERENCES

Creffield, J.W. (1991) *Wood Destroying Insects, Wood Borers and Termites.* CSIRO Australia Publication.

Edwards, R. and Mill, A.E. (1986) *Termites in Buildings.* Rentokil Limited, East Grinstead, 261 pp.

Ernst, E., Araujo, R.L. and TDRI (1986) In: Broughton, P. (ed.) *A Bibliography of Termite Literature* (1966–1978). John Wiley and Sons, Chichester, UK.

Grassé, P.P. (1982) *Termitologia,* Vol. I. Masson, Paris, 676 pp.

Grassé, P.P. (1984) *Termitologia,* Vol. II. Masson, Paris, 613 pp.

Grassé, P.P. (1986) *Termitologia,* Vol. III. Masson, Paris, 715 pp.

Haddlington, P. (1987) *Australian Termites and other Timber Pests.* Viliy Press, New South Wales.

Harris, W.V. (1969) *Termites as Pests of Crops and Trees.* Commonwealth Institute of Entomology, London, 41 pp.

Harris, W.V. (1971) *Termites: Their Recognition and Control,* 2nd edn. Longmans, London.

Hicken, N.E. (1971) *Termites: A World Problem.* Rentokil Library, Hutchinson, London, 232 pp. (revised by Edwards and Mill, see above.)

Howse, P.E. (1970) In: Cain A.J. (ed.) *Termites: A Study in Social Behaviour.* Hutchinson University Library, London, 150 pp.

Kofoid, C.A. (1934) *Termites and Termite Control.* 2nd edn. University of California Press, Berkeley, 830 pp.

Krishna, K. and Weesner, F.M. (eds) (1969) *Biology of Termites,* Vol. I Academic Press, New York, 598 pp.

Krishna, K. and Weesner, F. M. (eds) (1970) *Biology of Termites,* Vol. II. Academic Press, New York, 643 pp.

Lee, K.E. and Wood, T.G. (1971) *Termites and Soils.* Academic Press, New York.

Maeterlinck, M. (1927) *The Life of the White Ant.* George Allen and Unwin Ltd, 213 pp.

Marais, E.N. (1937) *The Soul of the White Ant.* Translated by W. de Kok. Methuen and Co. Ltd, London, 184 pp.

Noyes, H. (1937) *Man and the Termite*. Peter Davies Ltd, London, 289 pp.

Pearce, M.J. (1995) *The Termite Slide Kit. Termite Pests of Crops, Trees, Rangeland and Food Stores*. Natural Resources Institute Publication, 12 pp. (52 slides).

Rajagopal, D. (1990) Termite research in India. In: Veeresh, G.K., Kumar, A.R.V. and Shivashankar, T. (eds.) *Social Insects, an Indian Perspective*. IUSSI Indian Chapter, Bangalore, pp. 173–192.

Roonwal, M.L. (1979) *Termite Life and Termite Control in Tropical South Asia*. Jodhpur Scientific Publications, 177 pp.

Sands, W.A. (1965) A revision of the termite subfamily Nasutitermitinae (Isoptera: Termitidae) from the Ethiopian region. *Bulletin of the British Museum (Natural History)* Suppl.4, 1–172.

Sands, W.A. (1981) *The Social Life of Termites*. The Central Association of Bee Keepers, 16 pp.

Skaife, S.F. (1955) *Dwellers in Darkness: An Introduction to the Study of Termites*. Longmans, Green, London, 134 pp.

Snyder, T.E. (1956) Annotated subject, heading bibliography of termites. 1350 B.C. to A.D. 1954. *Smithsonian Miscellaneous Collections* 130, 1–305.

Snyder, T.E. (1961) Supplement to the annotated subject, heading bibliography of termites (1955–1960). *Smithsonian Miscellaneous Collections* 143, 1–137.

Snyder, T.E. (1968) Second supplement to the annotated subject, heading bibliography of termites (1961–1965). *Smithsonian Miscellaneous Collections* 152, 1–188.

Termite Abstracts (1980–1984) (Volumes 1–4) Centre for Overseas Pest Research, UK.

Termite Abstracts (1985–1988) (Volumes 5–8) Taylor and Francis Ltd, UK.

Termite Abstracts (1989–1992) (Volumes 9–11) Natural Resources Institute Chatham, UK.

Thorne, B.L. (1996) Kings and queens of the underworld. *Pest Control Technology* Part 1, 46–50, May 1996.

Wood, T.G. and Johnson, R.A. (1986) The biology, physiology, and ecology of termites. In: Vinson, S.B. (ed.) *Economic Impact and Control of Social Insects*. Praeger Publishers, New York, pp. 1–68.

Wilson, E.O. (1971) *The Insect Societies*. Harvard University Press, Cambridge, Massachusetts, USA, pp. 548.

TAXONOMIC REFERENCES

Ahmad, M. (1950) The phenology of termite genera based on imago, worker mandibles. *Bulletin of the American Museum of Natural History* 95, 37–86.

Akhtar, M.S. and Anwer R. (1991) Variability in the size of the soldier caste of the *Odontotermes obesus* (Rambur). *Pakistan Journal of Zoology* 20, 169–174.

Bacchus, S. (1987) A taxonomic and biometric study of the genus *Cryptotermes* (Isoptera: Kalotermitidae). *Tropical Pest Bulletin No.7*. Tropical Development and Research Institute Publication.

Bels, P.J. and Pataragetvit, S. (1982) Edible mushrooms in Thailand cultivated by termites. In: Chang, S.T. and Quimio, T.H. (eds) *Tropical Mushrooms*. Chinese University Press, Hong Kong, pp. 445–461.

Bouillon, A. and Mathot, G. (1965–1971) Quel est ce termite Africain? Part 1, *Zooleo* 1, 1–115 (1975); Part 2, *Zooleo* 1 (suppl. 1), 1–23 (1966); Part 3, *Zooleo* 1 (suppl. 2), 1–48 (1971).

Broughton, R.E. and Kistner, D.H. (1991) A DNA hybridisation study of the termite *Zootermopsis* (Isoptera: Termopsidae). *Sociobiology* 19, 15–40.

Chhotani, O.B. (1977) A review of taxonomy of Indian termites. *Record of the Zoolological Survey of India* 9, 1–36.

Coaton, W.G.H. and Sheasby, J.L. (1976) The genus *Coptotermes* Wasmann (Rhinotermitidae:Coptotermitinae). *National Survey of the Isoptera of Southern Africa 11. Cimbebasia* 3, 140–172.

Cowie, R.H. (1989) The zoogeographical composition and distribution of the Arabian termite fauna. *Biological Journal of the Linnean Society* 36, 157–168.

Cowie, R.H., Wood, T.G., Barnett, E., Sands, W.A. and Black, H.I.J. (1990) A check list of the termites of Ethiopia with a review of their biology, distribution and pest status. *African Journal of Ecology* 28, 21–23.

Disney, R.H.L. and Kistner, D.H. (1995) Revision of African Termitoxeniinae. *Sociobiology* 26, 117–225.

Goa, D.R., Lam, P.K.S. and Owen, P.T. (1992) The taxonomy, ecology and management of economically important termites in China. *Memoirs of the Hong Kong Natural History Society* 19, 15–50.

Harris, W.V. (1966) The genus *Ancistiotermes* (Isoptera). *Bulletin of the British Museum* (Natural History) 18, 3–20.

Harris, W.V. (1967) Termites of the genus *Anacanthotermes* in North Africa and Near East (Isoptera: Hodotermitidae). *Proceedings of the Royal Entomological Society of London* B 36, 79–86

Harris, W.V. (1968) Termites of the Sudan. *Sudan Natural History Museum Bulletin* 4, 1–29.

Haverty, M.I.L., Nelson, J. and Page, M. (1991) Preliminary investigations of the cuticular hydrocarbons from North American *Reticulitermes* and tropical and subtropical *Coptotermes* (Isoptera: Rhinotermitidae) for chemotaxonomic studies. *Sociobiology* 19, 51–76.

Huang fu, Sheng, Li Gui, Xiang and Zhu Shimo (1989) *The Taxonomy and Biology of Chinese Termites.* Tianze Press, Guanghou, China, 605 pp.

Jarzembowski, E.A. (1981) An early cretaceous termite from southern England. (Isoptera: Hodotermitidae). *Systematic Entomology* 6, 91–96.

Johnson, R.A. (1979) Configuration of the digestive tube as an aid to identification of worker Termitidae (Isoptera). *Sytematic Entomology* 4, 31–38.

Johnson, R.A., Lamb, R.W., Sands, W.A., Shittu, M.O., Williams, R.M.C. and Wood, T.G. (1978) A check list of Nigerian termites (Isoptera) with brief notes on their biology and distribution. *Nigerian Field* 45, 50–64.

Korman, A.K. and Pashley, D.P. (1991) Genetic comparisons among US populations of Formosan subterranean termites. *Sociobiology* 19, 41–50.

Krishna, K. (1961) A generic revision and phylogenetic study of the family Kalotermitidae (Isoptera). *Bulletin of the American Museum of Natural History* 122, 303–408.

Krishna, K. (1970) Taxonomy, phylogeny and distribution of termites. In: Krishna, K. and Weesner, F.M. (eds) *Biology of Termites Vol. 2.* Academic Press, New York, pp. 127–152.

Krishna, K. and Emerson, A.E. (1983). A new fossil species of termite from Mexican amber, *Mastotermes electromexicus* (Isoptera: Mastotermitidae). *American Museum Novitiates* 2767, 1–8.

Mathews, A.G.A. (1977) *Studies on termites from the Mato Grosso State, Brazil.* Rio de Janeiro, Academia Brasiliera der Ciencias.

Mitchell, B.L. (1980) Report on a survey of termites of Zimbabwe. In: *Occasional Papers of the National Museums and Monuments,* Series B. Natural Sciences 6, 323 pp.

Nel, A. and Arillo, A. (1995) Revision of *Macrotermes haidingeri* (Heir 1849) and description of two new fossil *Mastotermes* from the Oligocene of France and Spain (Isoptera: Mastotermitidae). *Bulletin de la Societe Entomologique de France* 100, 67–74.

Pearce, M.J. (1993) *What Termite? A Guide to Identification of Termite Pest Genera in Africa.* Natural Resources Institute Publication, Technical Leaflet No. 4, 19 pp.

Pearce, M.J. and Waite, B.S. (1994) A list of termite genera (Isoptera) with comments on taxonomic changes and regional distribution. *Sociobiology* 23, 247–263.

Pearce, M.J., Tiben, A., Kambal, M.A., Thomas, R.J. and Wood, T.G. (1986) Termites (Isoptera) from the Tokar delta and Red Sea coastal areas of Sudan. *Journal of Arid Environments* 10, 193–197.

Roonwal, M.L. (1969) Measurement of termites (Isoptera) for taxonomic purposes. *Journal of the Zoological Society of India* 21, 9–66.

Roonwal, M.L. and Chotani, O.B. (1989) *The Fauna of India and Adjacent Countries* (Isoptera: Termites). Vol. 1. Zoological Survey of India, Calcutta, 664 pp.

Ruelle, J.E. (1970) A revision of the termites of the genus *Macrotermes* from the Ethiopian region (Isoptera: Termitidae). *Bulletin of the British Museum of Natural History* 34, 365–444.

Sands, W.A. (1969) The association of termites and fungi. In: Krishna, K. and Weesner, F.M. (eds) *Biology of Termites* Vol. 1. Academic Press, New York. pp. 495–524.

Sands, W.A. (1972) The soldierless termites of Africa (Isoptera: Termitidae). *Bulletin of the British Museum of Natural History* 18, 1–244.

Sands, W.A. (1992) The termite genus *Amitermes* in Africa and the Middle East. *Natural Resources Bulletin* 51, 140 pp. (Also on computer key on disk.)

Sands, W.A. (1995) New genera and species of soil feeding termites (Isoptera: Termitidae) from African savannas. *Journal of Natural History* 29, 1483–1515.

Sands, W.A (1997). *Genera Key to African Termites Based on the Worker Caste.* CAB International, Wallingford (in press).

Snyder, T.E. (1949) Catalogue of the termites of the world. *Smithsonian Miscellaneous Collections* 112, 1–490.

Watson, J.A.L. and Abbey H.M. (1993) *An Atlas of Australian Termites.* CSIRO Publication, Southern Canberra.

Webb, G.C. (1961) *Keys to the Genera of African Termites.* Ibadan University Press, Nigeria, 36 pp.

Weesner, F.M. (1965) *The Termites of the United States, a Handbook.* National Pest Control Association, New Jersey, 70 pp.

Williams, R.M.C. (1966) The east African termites of the genus *Cubitermes* (Isoptera: Termitidae). *Transactions of the Royal Entomological Society London* 118, 73–118.

REFERENCES ON BIOLOGY AND CONTROL

Abe, T. (1987) Evolution of life types in termites. In: Kawano S., Connell J.H. and Hidaka T. (eds) *Evolution and Coadaptation in Biotic Communities.* University of Tokyo Press, Tokyo, pp. 125–148.

Abushama, F.T. and Kambal, M.A. (1977a) The role of sugars in the food selection of the termite *Microtermes tragardhi* (Sjöstedt). *Zeitschrift für Angewandte Entomologie* 84, 250–255.

Abushama, F.T. and Kambal, M.A. (1977b) Field observations on attack of sugarcane by the termite *Microtermes tragardhi* (Sjöstedt). *Zeitschrift für Angewandte Emtomologie* 82, 355–359.

Agarwal, V.B. (1978) Physical and chemical constituents of fungus combs of *Odontotermes microdentatus* Roonwal and Sen-Sarma and *Odontotermes obesus* Rambur (Isoptera: Termitidae). *Bulletin of the Zoological Survey of India* 1, 15–19.

Akhtar, M.S. and Bhatti, M. (1993) Studies on the toxicity of Karate, sumicidin and arrivo against *Microcerotermes championi* (Isoptera: Termitidae). *Sociobiology* 23, 115–125.

Akhtar, M.S. and Saleem, M. (1993) Toxicity of insecticides against *Coptotermes heimi. Pakistan Journal of Zoology* 25, 139–142.

Akhtar, M.S. and Shahid, A.S. (1990) Termite (Isoptera) population and damage in sugarcane fields at Manga mandi, Lahore, Pakistan. *Proceedings of Pakistan Congress of Zoology* 1043–1053.

Akhtar, M.S. and Shahid, A.S. (1991) Efficacy of chlorpyrifos and dieldrin in cotton fields against subterranean termites. *Pakistan Journal of Zoology* 23, 133–137.

Akhtar, M.S. and Shahid, A.S. (1993) Termites as pests of agricultural crops in Pakistan. *Pakistan Journal of Zoology* 25, 187–193.

Alderton, D. (1992) *Parrots.* Whittet Books Ltd, London, 127 pp.

Al Fazairy, A. and Ahlam, A. (1987) Infection of termites by *Spodoptera littoralis* nuclear polyhedrosis virus. *Insect Science and its Application* 9, 37–39.

Amburgey, T.L. and Smythe, R.V. (1977) Factors influencing termite feeding on brown rotted wood. *Sociobiology* 3, 3–12.

Ampion, A. and Quennedey, A. (1981) The abdominal epidermal glands of termites and their phylogenetic significance. In: Howse, P.E. and Clement, J.L. (eds) *Biosystematics of Social Insects.* Academic Press, London, pp. 249–261.

Bach, C. (1990) First international symposium on termite management in historic buildings. *Sociobiology* 17, 218 pp.

Bacchus, S. (1979) New exocrine gland on the legs of some Rhinotermitidae (Isoptera). *International Journal of Insect Morphology and Embryology* 8, 135–142.

Badertscher, S., Gerber, C. and Leuthold, R.H. (1983) Polyethism, food supply and processing in termite colonies of *Macrotermes subhyalinus. Behavioural Ecology and Sociobiology* 12, 115–119.

Becker, G. (1969) Rearing of termites and testing methods used in the laboratory. In: Khrishna, K. and Weesner, F.M. (eds) *Biology of Termites* Vol. 1. Academic Press, New York, pp. 351–385.

Becker, G. (1974) The orientation and feeding of termites. *Sitzungsberichte de Gesellscaft Naturforschender Freunde Berlin* 14, 43–44.

Becker, G. (1976) Reaction of termites to weak magnetic fields. *Naturwissenschaften* 63, 201–202.

Berry, R.W. (1994) *Termites and Tropical Building.* UK Building Research Establishment (OBN201), March 1994.

Black, H.I. and Wood, T.G. (1989) Effect of cultivation on vertical distribution of *Microtermes* spp. (Isoptera: Termitidae: Macrotermitinae) in soil at Mokwa, Nigeria. *Sociobiology* 15, 133–138.

Boicca, A.L., Jr, Matioli, J.C., Fazolin, M., Nakano, O. and Rivero, R.C. (1992) Diffusion of phosphine inside nests of *Cornitermes cumulans* (Kollar) (Isoptera, Termitidae). *Anais da Sociedade Entomologica do Brasil* 21, 307–317.

Bordereau, C., Van Tuyen, V. and Robert, A. (1994) Self sacrifice in *Globitermes sulphureus* Haviland soldiers, frontal gland dehiscence. In: Lenoir, A., Arnold, G. and Lepage, M. (eds) *Exocrine Glands and Social Organisation.* Les Insectes Sociaux, Paris, Sorbonne, August 1994, Abstracts 12th International Congress International Union for the Study of Social Insects, p. 225.

Breznak, J.A. (1983) Biochemical aspects of symbiosis between termites and their intestinal microbiota. In: Anderson, J.M., Rayner, A.D. and Walton, D.H. (eds) *Invertebrate Microbial Interactions.* Cambridge University Press, Cambridge, pp. 173–203.

British Standards Institute (1989) *Wood preservatives. Determination of toxic values against* Reticulitermes santonensis *de Feytaud (laboratory method*). BS6239 (1990). British Standards Institution Publication December 1989. European Standard 117, 10 pp.

British Standards Institute (1990) *Wood preservatives. Determination of preventative action against* Reticulitermes santonensis *de Feytaud (laboratory method).* BS6240 (1990). British Standards Institution Publication March, European standard 118, 8 pp.

Cabrera, B.J. and Rust, M.K. (1994) Effect of temperature and relative humidity on the survival and wood consumption of the western drywood termite *Incisitermes minor* (Isoptera: Kalotermitidae). *Sociobiology* 24, 95–113.

Castle, G.B. (1934) In: *The Dampwood Termites of Western United States, Genus* Zootermopsis, pp. 273–291.

Clement, J.L., Jequel., M., Leca, J.L., Lohou, C .and Burban, G. (1996) Elimination of foraging populations of *Reticulitermes santonenesis* in one street of Paris, France, using hexaflumuron baits. In: Wildey, K.B. (ed.) *Proceedings of the International Conference on Insect Pests in the Urban Environment.* Edinburgh, UK, 640 pp.

Collins, N.M. (1979a) The nests of *Macrotermes bellicosus* (Smeathman) from Mokwa, Nigeria. *Insectes Sociaux* 26, 240–246.

Collins, N.M. (1979b) Observations on foraging activity of *Hospitalitermes umbrinus* (Haviland) Isoptera, Termitidae in the Gunong Mulu National Park Sarawak. *Ecological Entomology* 4, 231–238.

Connors, J. (1963) *The Miami Herald*, 6 November 1963, pp. 2–6.

Cowie, R.H., Logan, J.W.M. and Wood, T.G. (1989) Termite (Isoptera) damage and control in tropical forestry with special reference to Africa and Indo-Malaysia, a review. *Bulletin of Entomological Research* 79, 173–184.

Curtis, D.A. and Waller, A.W. (1995) Changes in nitrogen fixation rates in termites (Isoptera: Rhinotermitidae) maintained in the laboratory. *Annals of the Entomological Society of America* 86, 764–767.

Danthanarayana, W. and Vitarana, I. (1987) Control of the livewood tea termite *Glyptotermes dilatatus* using *Heterorhabditis* sp. (Nemat.). *Agriculture, Ecosystems and Environment* 19, 333–342.

Darlington, J.P.E.C. (1985a) The structure of mature mounds of the termite *Macrotermes michaelseni* in Kenya. *Insect Science and its Application* 6, 149–156.

Darlington, J.P.E.C. (1985b) Multiple primary reproductives in the termite *Macrotermes michaelseni* (Sjöstedt). In: Watson, J.A.L., Okot-Kotber, B.M. and Noirot, C. (eds) *Caste Differentiation in Social Insects. Current Theories in Tropical Science* Vol. 3. Pergamon Press, Oxford, pp. 187–200.

Darlington, J.P.E.C. (1987) How termites keep cool. *Entomological Society of Queensland New Bulletin* 15, 45–46.

Darlington, J.P.E.C. (1994) Nutrition and evolution in fungus growing termites. In: Hunt, J.H. and Nalepa, C.A. (eds) *Nourishment and Evolution in Insect Societies*. Westview Press, Boulder, pp. 105–130.

Dejean, A. and Ruelle, J.E. (1995) Importance of *Cubitermes* termitaries as shelter for alien termite societies. *Insectes Sociaux* 42, 129–136.

Delate, K.M. and Grace, J.K. (1995a) Potential use of pathogenic fungi in baits to control the Formosan subterranean termite, *Coptotermes formosanus* Shiraki (Isoptera: Rhinotermitidae). *Journal of Applied Entomology* 119, 429–433.

Delate, K.M. and Grace, J.K. (1995b) Susceptibility of neem to attack by the Formosan subterranean termite, *Coptotermes formosanus* Shiraki (Isoptera: Rhinotermitidae). *Journal of Applied Entomology* 119, 93–95.

Delate, K.M., Grace, J.K., Armstrong, J.W. and Tome, C.H.M. (1995) Carbon dioxide as a potential fumigant for termite control. *Pesticide Science* 44, 357–361.

Deligne, J., Quennedey, A. and Blum, M.S. (1981) The enemies and defence mechanisms of termites. In: Hermann, H.R. (ed.) *Social Insects*. Vol. 2. Academic Press, New York, pp. 1–76.

de Wilde, J. and Beetsma, J. (1982) The physiology of caste development in social insects. *Advances in Insect Physiology* 16, 167–246.

Dhanarajan, G. (1978) Cannibalism and necrophagy in a subterranean termite (*Reticulitermes lucifugus* var. *santonensis*. *Malayan Nature Journal* 31, 237–251.

Disney, R.H.C. (1994) *Scuttle Flies: The Horidae*. Chapman and Hall, London, 467 pp.

Duelli, P. and Duelli-Klein, R. (1978) The magnetic effect on termite mound orientation in *Amitermes meridionalis*. *Mitteilungen de schweizenschen Entomologischen Gesellschaft* 51, 337–342.

Eggleton, P., Bignell, D.E., Sands, W.A., Waite, B., Wood, T.G. and Lawton, J.H. (1995) The species richness of termites (Isoptera) under differing levels of forest disturbance in the Mbalmayo Forest Reserve, southern Cameroon. *Journal of Tropical Ecology* 11, 85–98.

Eisner, T., Kriston, I. and Aneshansley, D.J. (1976) Defensive behaviour of a termite *Nasutitermes exitiosus. Behavioral Ecology and Sociobiology* 1, 83–126.

El Bakri, A., Eldein, N., Kambal, M.A., Thomas, R.J. and Wood, T.G. (1989a) Colony foundation and development in *Microtermes* sp. nr. *albopartitus* (Isoptera: Macrotermitinae) in Sudan. *Sociobiology* 15, 169–173.

El Bakri, A., Eldein, N., Kambal, M.A., Thomas, R.J. and Wood, T.G. (1989b) Effect of fungicide impregnated food on the viability of fungus combs and colonies of *Microtermes* sp. nr. *albopartitus* (Isoptera: Macrotermitinae). *Sociobiology* 15, 175–180.

Emerson, A.E. (1955) Geographical origins and dispersion of termite genera. *Fieldiana Zoolology* 37, 465–521.

Epsky, D. and Capinera, L.J. (1988) Efficiency of the entomogenous nematode *Steiernema feltiae* against a subterranean termite, *Reticulitermes tibialis* (Isoptera: Rhinotermitidae). *Journal of Economic Entomology* 81, 1313–1317.

Ettershank, G., Ettershank, J.A. and Whitford, W.G. (1980) Location of food resources by subterranean termites. *Environmental Entomology* 9, 645–648.

Fajairj, A.A. (1987) Infection of termites by *Spodoptera littoralis* nuclear polyhedrosis virus. *Insect Science and its Application* 9, 37–39.

Fernandes, P.M. and Alves, S.B. (1991) Control of *Cornitermes cumulanus* (Kollar, 1832) (Isoptera: Termitidae) with *Beauveria bassiana* (Bals) Vuill and *Metarhizium anisopliae* (Metsch) Sorok under field conditions. *Annais da Sociedade Entomologia do Brasil* 20, 45–49.

Ferrar, P. and Watson, J.A.L. (1970) Termites associated with dung in Australia. *Journal of the Australian Entomological Society* 9, 100–102.

Forschler, B.T. (1994) Fluorescent spray paint as a topical marker on subterranean termites (Isoptera: Rhinotermitidae). *Sociobiology* 24, 27–38.

French, J.R.J. (1993) Physical barriers and bait toxicants in future termite control strategies. *Annals of Entomology* 9, 1–5.

French, J.R.J., Rasmussen, R.A., Ewart, D.M. and Khalil, M.A.K. (1977) The gaseous environment of mound colonies of the subterranean termite *Coptotermes lacteus* (Isoptera: Rhinotermitidae) before and after feeding on mirex treated decayed wood bait blocks. *Bulletin of Entomological Research* 87, 145–149.

Gold, R.E., Howell, J.R.H.N., Pawson, B.M., Wright, M.S. and Lutz, J.C. (1996) Evaluation of termiticyde residues and bioavailability from five soil types and locations in Texas. In: Wildey, K.B. (ed.) *Proceedings of the International Conference on Insect Pests in the Urban Environment.* Edinburgh, UK, pp.467–484.

Grace, J.K. (1989) Northern subterranean termites. *Pest Management* 8, 14–16.

Grace, J.K. (1990) *Termites in Eastern Canada: An Updated Review and Bibliography.* The International Research Group on Wood Preservation Stockholm, Sweden Doc. IRG/WP/1431 6 pp.

Grace, J.K. (1991) Response of eastern subterranean termites (Isoptera, Rhinotermitidae) to borate dust and soil treatments. *Journal of Economic Entomology* 84, 1753–1757.

Grace, J.K. (1992) Termite distribution, colony size and potential for damage. *Proceedings of the National Conference on Urban Entomology,* pp. 67–76.

Grace, J.K. (1994) Protocol for testing effects of microbial pest control agents on non target subterranean termites (Isoptera: Rhinotermitidae). *Journal of Economic Entomology* 87, 269–274

Grace, J.K. and Ewart, D.M. (1996) Recombinant cells of *Pseudomonas fluorescens:* a highly palatable encapsulation for delivery of engineered toxins to subterranean termites (Isoptera: Rhinotermitidae). *Letters in Applied Microbiology* 23, 183–186.

Grace, J.K. and Yamamoto, T. (1993) Diatomaceous earth is not a barrier to formosan subterranean termites (Isoptera: Rhinotermitidae). *Sociobiology* 23, 25–30.

Grace, J.K. and Yates, J.R. (1992) Behavioural effects of neem insecticide on *Coptotermes formosanus* (Isoptera: Rhinotermitidae). *Tropical Pest Management* 38, 176–180.

Grace, J.K. and Zoberi, M.H. (1992) Experimental evidence for transmission of *Beauveria bassiana* by *Reticulitermes flavipes* workers (Isoptera: Rhinotermitidae). *Sociobiology* 20, 23–28.

Grace, J.K., Wood, D.L. and Frankie, G.W. (1988) Trail following behaviour of *Reticulitermes hesperus* Banks (Isoptera: Rhinotermitidae). *Journal of Chemical Ecology* 14, 653–667.

Grace, J.K., Abdallay, A. and Farr, K.R. (1989) Eastern subterranean termite (Isoptera: Rhinotermitidae) foraging territories and populations in Toronto. *Canadian Entomologist* 121, 551–556.

Grace, J.K., Yamamoto, R.T. and Laks, P.E. (1993) Evaluation of the termite resistance of wood pressure treated with copper naphthenate. *Forest Products Journal* 43, 72–75.

Grassé, P.P. (1967) New experiments with *Macrotermes mülleri* and considerations on the theory of stigmergy. *Insectes Sociaux* 14, 73–102.

Gui, Xiang, Zi, L., Rong, D. and Biao, Y. (1994) Introduction to termite research in China. *Journal of Applied Entomology* 117, 360–369.

Hanel, H. and Watson, J.A.L. (1983) Preliminary field tests on the use of *M. anisoplea* for the control of *Nasutitermes exitiosus* (Hill) (Isoptera: Termitidae). *Bulletin of Entomological Research* 73, 305–313.

Jander, R. and Daumer, K. (1974) Guideline and gravity orientation of blind termites foraging in the open (Termitidae, *Macrotermes, Hospitalitermes*). *Insectes Sociaux* 21, 45–69.

Johnson, R.A., Lamb, R.W. and Wood, T.G. (1981a) Termite damage and crop loss studies in Nigeria: a survey of damage to groundnuts. *Tropical Pest Management* 27, 325–342.

Johnson, R.A., Thomas, R.J., Wood, T.G. and Swift, M.J. (1981b) The inoculation of the fungus comb in newly founded colonies of some species of the Macrotermitinae (Isoptera). *Nigeria Journal of Natural History* 15, 751-756.

Jones, S.C. (1988) Foraging and distributions of subterranean termites. *Proceedings of the National Conference on Urban Entomology,* pp. 23–32.

Jones, S.C. (1991) Field evaluation of boron as a bait toxicant for *Heterotermes aureus* (Isoptera: Rhinotermitidae). *Sociobiology* 19, 187–209.

Jones, W.E., Grace, J.K. and Tamashiro, M. (1996) Virulence of seven isolates of *Beauveria bassiana* and *Metarhizium anisopliae* to *Coptotermes formosanus* (Isoptera: Rhinotermitidae). *Environmental Entomology* 25, 481–487.

Kettler, R. and Leuthold, R.H. (1995) Inter and intraspecific alarm response in the termite *Macrotermes subhyalinus* (Rambur). *Insectes Sociaux* 42, 145–156.

Khan, M.A. (1980) Effect of relative humidity on termites under starvation conditions. *Zeitschrift für Angewandte Zoologie* 67, 133–178.

Khan, M.A. (1980) Effect of humidity on feeding activities and development of termites. *Material und Organismen* 15, 167–194.

Khan, K.I., Jaffri, R.H. and Ahmad, M. (1985) The pathogenicity and development of *Bacillus thuringiensis* in termites. *Pakistan Journal of Zoology* 17, 201–209.

Kirchner, W.H., Broecker, I. and Tautz, J. (1994) Vibrational alarm communication in the dampwood termite *Zootermopsis nevadensis. Physiological Entomology* 19, 187–190.

Kruuk, H. and Sands, W.A. (1972) The aardwolf (*Proteles cristatus* Sparrman) 1783 as predator of termites. *East African Wildlife Journal* 10, 211–227.

Lai, P.Y., Tamashiro, M. and Fujii, J.K. (1982) Pathogenicity of six strains of entomogenous fungi to *Coptotermes formosanus. Journal of Invertebrate Pathology* 39, 1–5.

Lenz, M. (1994) Food resources, colony growth and caste development in wood feeding termites. Chapter 6. In: Hunt, J.H. and Nalepa, C.A. (eds) *Nourishment and Evolution in Insect Societies.* Westview Press, Boulder, pp. 159–209.

Lenz, M. and Runko, S. (1994) Protection of buildings, other structures and material in ground contact from attack by subterranean termites (Isoptera) with a physical barrier, a fine mesh of high grade stainless steel. *Sociobiology* 24, 1–16.

Lenz, M., Amburgey, T.L., Zi, Rong, Kuhne, H., Maudlin, J.K., Preston, A.F. and Westcott, M. (1987) Inter-laboratory studies on termite, wood decay fungi associations, 1. Determination of maintenance conditions for several species of termites (Isoptera: Mastotermitidae, Termopsidae, Rhinotermitidae, Termitidae). *Sociobiology* 13, 1–56.

Lepage, M., Morel, G. and Resplendino, C. (1974) Discovery of termite galleries extending deep underground to water level in north Senegal. *Comptes rendus Hebdomadaires des Séances de l'Académie des Sciences* (Série d) 278, 1855–1858.

Li-G.X., Dai, Z. Rong and Yang, Z.B. (1994) Introduction to termite research in China. *Journal of Applied Entomology* 117, 360–369.

Logan, J.W.M. (1992) Termites (Isoptera) a pest or resource for small farmers in Africa. *Tropical Science* 32, 71–77.

Logan, J.W.M. (1995) Termites (Isoptera). In: Matthews, G.A. and Tunstall, J. (eds) *Insect Pests of Cotton.* CAB International, Wallingford, UK.

Logan, J.W.M. and Buckley, D.S. (1991) Subterranean termite control in buildings. *Pesticide Outlook* 2, 33–37.

Logan, J.W.M., Cowie, R.H. and Wood, T.G. (1990) Non chemical control in crops and forestry, a review. *Bulletin of Entomological Research* 80, 309–330.

Logan, J.W.M., Rajagopal, D., Wightman, J.A. and Pearce, M.J. (1992) Control of termites and other pests of groundnuts with special reference to controlled release formulations of non persistent insecticides. *Bulletin of Entomological Research* 82, 57–66.

Lüscher, M. (1972) Environmental control of juvenile hormone (JH) secretion and caste differentiation in termites. *General and Comparative Endocrinology* 19, 509–514.

Lys, J.A. and Leuthold, R.H. (1991) Task specific distribution of the two worker castes in extranidial activities in *Macrotermes bellicosus* (Smeathman), observation of behaviour during food acquisition. *Insectes Sociaux* 38, 161–170.

Lys, J.A. and Leuthold, R.H. (1994) Forces affecting imbibition in *Macrotermes* workers (Termitidae: Isoptera). *Insectes Sociaux* 41, 79–84.

Maistrello, M. and Sbrenna, G. (1994) Behavioural profiles in laboratory colonies of *Kalotermes flavicollis* (Isoptera: Kalotermitidae). Poster presentation. Les Insectes Sociaux, Paris, Sorbonne, August 1994. In: Lenoir, A., Arnold, G. and Lepage, M. (eds) *Abstracts 12th International Congress International Union for the Study of Social Insects.*

Mampe, C.D. (1990) Termites. In: Mallis, A. (ed.) *Handbook of Pest Control. The Behaviour, Life History and Control of Household Pests.* Franyak and Foster, Cleveland, Ohio, pp. 201–263.

Mampe, C.D. (1997) What does the crystal ball say about the future of termite control? *Pest Control* 65, 64–65.

Martius, C. (1994) Diversity and ecology of termites in Amazonian forests. *Pedobiologia* 38, 407–428.

Martius, C., Wassmann, R., Thein, U., Bandeira, A., Rennenberg, H., Junk, W. and Seiler, W. (1993) Methane emission from wood feeding termites in Amazonia. *Chemosphere* 26, 623–632.

Maudlin, K.J. and Beal, H.R. (1989) Entomogenous nematodes for control of subterranean termites *Reticulitermes* spp. (Isoptera: Rhinotermitidae). *Journal of Economic Entomology* 82, 1638–1645.

McKevan, D.K. (1978) *Land of the Locusts.* Part 1. Lyman Entomological Museum and Research Laboratory Memoir No.6 (Special publication) No.14, Canada, 529 pp.

Mitchell, M.R. (1989) Susceptibility to termite attack of various tree species planted in Zimbabwe. In: Roland, D.J. (ed.) *Trees for the Tropics.* Australian Centre for International Agricultural Research, Monograph 10, pp. 215–227.

Moore, H. (1993) Using Tim-Bor to control wood destroying organisms. *Pest Control Magazine* 61, February.

Myles, T.G. (1994) Use of disodium octoborate tetrahydrate to protect aspen waferboard from termites. *Forest Products Journal* 44, 33–36.

Myles, T.G. (1995) New records of drywood termites, introduction, interception and extirpation. *Proceedings of the Entomological Society of Ontario* 726, 77–83.

Myles, T.G. Abdallay, A. and Sisson, J. (1994) 21st century termite control. *Pest Control Technology* March 1994, 64–68, 70, 72, 108.

Nkunika, P.O. (1992) Termites as a major pest in Zambia. Prospects for their control. In: *Proceedings of the 1st Regional Workshop on Termite Control.* Nairobi, Kenya, 17–19 August, 1992.

Noirot, C. (1995) The sternal glands of termites, segmental pattern, phylogenetic implications. *Insectes Sociaux* 42, 321–323.

Noirot, C. and Noirot-Timothee, C. (1969) The digestive system. In: Krishna, K. and Weesner, F.M. (eds) *Biology of Termites.* Academic Press, New York, pp. 49–88.

Nutting, W.L. and Jones, S.C. (1990) Methods for studying the ecology of subterranean termites. *Sociobiology* 17, 167–189.

Oi, F.M. and Su, N.Y. (1994) Stains tested for marking *Reticulitermes flavipes* and *R. virginicus* (Isoptera: Rhinotermitidae). *Sociobiology* 24, 241–248.

Okwakol, M.J.N (1991) Fauna associated with *Cubitermes testaceus* mounds (Isoptera: Termitidae). *Insect Science and its Application* 12, 511–514.

Omo-Malaka, S.L. (1972) Some measures applied in the control of termites in parts of Nigeria. *Nigerian Entomology Magazine* 2, 137–141.

Pearce, M.J. (1987) *Antennopsis gayi* Buchli parasitises a termite species in Domoga Bone National Park, Sulawesi. *Antenna* 11(3), 89.

Pearce, M.J. (1988) Seals, tombs, mummies and tunnelling in the drywood termite *Cryptotermes* (Isoptera: Kalotermitidae). *Sociobiology* 13, 217–226.

Pearce, M.J. (1990). A new trap for collecting termites and assessing their foraging activity. *Tropical Pest Management* 36, 310–311

Pearce, M.J., Cowie, R.H., Pack, A.S. and Reavey, D. (1990) Intraspecific aggression, colony identity and foraging distances in Sudanese *Microtermes* spp. (Isoptera, Termitidae, Macrotermitinae). *Ecological Entomology* 15, 71–77.

Pearce, M.J., Waite, B.S., Evans, R.A. and Logan, J.W.M. (1991) A large vertical glass plate for observing the behaviour of *Microtermes* sp. nr. *lepidus* (Isoptera: Macrotermitinae) and other subterranean termites. *Sociobiology* 19, 323–331.

Percy, J.R. and Weatherston, J. (1974) Gland structure and pheromone production in insects. In: Birch, M.C. (ed.) *Pheromones.* North Holland Publishing Company, Amsterdam, pp. 12–34.

Prestwich, G.D. (1982) The Total Termite. *The Science Teacher* 49, 4 April, 3 pp.

Prestwich, G.D. (1983) The chemical defences of termites. *Scientific American* 249(1), 68–75.

Prestwich, G.D. (1984) Defence mechanisms of termites. *Annual Review of Entomology* 29, 201–232.

Quennedey, A. (1975) Morphology of exocrine glands producing pheromones and defensive secretions in sub-social and social insects. In: Noirot, C.H., Howse, P.E. and Le Manse, G. (eds) *Pheromones and Defensive Secretions in Social Insects.* IUSSI Publication, Dijon, pp. 1–21.

Rajagopal, D. (1984) Observations on the natural enemies of *Odontermes wallonensis* (Wasmann, Isoptera: Termitidae) in south India. *Journal of Soil Biology and Ecology* 4, 102–107.

Rajagopal, D. (1985) Population estimation and seasonal fluctuations of mound building termite *Odontotermes wallonensis* (Isoptera: Termitidae) in south India. *Sociobiology* 11, 67–76.

Rajagopal, D. (1987) Termite research in India. In: Veeresh, G.K., Kumar, A.R.V. and Shivashankar, I. (eds) *Social Insects: An Indian Perspective* – IVSSI, Bangalore, pp. 173–192.

Richard, G. (1969) Nervous system and sense organs. In: Krishna, K. and Weesner, F.M. (eds) *Biology of Termites.* Vol. 1. Academic Press, New York, pp. 161–192.

Robinson, W.H. (1994) Producing and applying termiticide foam. *Pest Management* 13, 20–22.

Salih, A.G.M. and Logan, J.W.M. (1990) Histological dyes for marking *Microtermes lepidus* (Isoptera: Macrotermitinae). *Sociobiology* 16, 247–250.

Salik, J. and Tho Yow Pong (1984) An analysis of termite fauna in Malayan rain forest. *Journal of Applied Ecology* 21, 547–561.

Salman, A.G.A. and Sayed, A.A. (1990) Construction techniques for the control of the sand termite *Psammotermes hybostoma* Desneux in traditional housing in Egypt. *Tropical Pest Management* 36, 68–70.

Sands, W.A. (1960) The initiation of fungus comb construction in laboratory colonies of *Ancistrotermes guineesis* (Silvestri). *Insectes Sociaux* 7, 251–263.

Sands, W.A. (1965) Alate development and colony foundation in five species of *Trinervitermes* (Isoptera: Nasutitermitinae) in Nigeria, W. Africa. *Insectes Sociaux* 12, 117–130.

Sands, W.A. (1982) Agonistic behaviour of African soldierless Apicotermitinae (Isoptera: Termitidae). *Sociobiology* 7, 61–72.

Sasaki, K., Ishizaka, T., Suzuki, T., Takeda, M. and Uchiyama, M. (1991) Organonochlorine chemicals in skin lipids as an index of their accumulation in the human body. *Environmental Contamination and Toxicology* 21, 190–194.

Schoknecht, U., Rudolph, D. and Hertel, H. (1994) Termite control with micro-encapsulated permethrin. *Pesticide Science* 40, 49–55.

Scott Turner, J. (1994) Ventilation and thermal constancy of a colony of a southern African termite (*Odontotermes transvaalensis*, Macrotermitinae). *Journal of Arid Environments* 8, 231–248.

Sen Sarma, P.K. (1986) Economically important termites and their management in the oriental region. In: Vinson, S.B. (ed.) *Economic Impact and Control of Social Insects.* Praeger Publishers, New York, pp. 69–102.

Serment, M.M. and Pruvost, A.M. (1991) Termites in France, geographical distribution, spread, prevention. *CTBA, Information* 36, 35–38.

Sieber, R. and Leuthold, R.H. (1981) Behavioural elements and their meaning in incipient colonies of the fungus growing termite *Macrotermes michaelseni* (Isoptera: Macrotermitinae). *Insectes Sociaux* 28, 371–382.

Slaytor, M. (1994) Cellulose digestion in termites and cockroaches, What role do symbionts play? *Comparative Entomology* 107, 1–10.

Smith, J.L. and Rust, M.K. (1994) Termite preferences of the western subterranean termite, *Reticulitermes hesperus* Banks. *Journal of Arid Environments* 28, 313–323.

Sponsler, R.C. and Appel, A.G. (1991) Temperature tolerance of formosan and eastern subterranean termites. *Journal of Thermobiology* 16, 41–44.

Staples, J. and Milner, R. (1996) *Metarhizium anisopliae* as a mycotermiticide. Laboratory behavioural bioassay. In: Wildey, K.B. (ed.) *Proceedings of the International Conference on Insect Pests in the Urban Environment.* Edinburgh, UK, p. 493.

Steward, R.C. (1982) Comparison of the behavioural and physiological responses to humidity of five species of the dry, wood termites, *Cryptotermes* species. *Physiological Entomology* 7, 71–82.

Steward, R.C. (1983) Microclimate and colony foundation by imago and neotenic reproductives of dry wood termite species (*Cryptotermes* sp.) (Isoptera: Kalotermitidae). *Sociobiology* 7, 311–332.

Storey, G. (1995) Imidacloprid, a developing new termiticide chemistry. *Pest Control Magazine* February, pp. 36–37.

Su, N.Y. (1984) Alate production of a field colony of the formosan subterranean termite (Isoptera: Rhinotermitidae). *Sociobiology* 13, 167–172.

Su, N.Y. (1990) Measuring termiticides. *Pest Control Magazine* 58, 24, 30, 34, 35.

Su, N.Y. (1994a) The termite bait age dawns. *Pest Control Magazine* 36, June.

Su, N.Y. (1994b) Field evaluation of a hexaflumuron bait for population suppression of subterranean termites (Isoptera: Rhinotermitidae). *Journal of Economic Entomology* 87, 389–397.

Su, N.Y. and Scheffrahn, R.H. (1987) Alate production of a field colony of the formosan subterranean termite (Isoptera: Rhinotermitidae). *Sociobiology* 18, 167–172.

Su, N.Y. and Scheffrahn, R.H. (1988) The formosan termite. *Pest Management Magazine* July, pp.16–25.

Su, N.Y. and Scheffrahn, R.H. (1990a) Economically important termites in the United States and their control. *Sociobiology* 17, 77–94.

Su, N.Y. and Scheffrahn, R.H. (1990b) Potential of insect growth regulators. *Sociobiology* 17, 313–328.

Su, N.Y. and Scheffrahn, R.H. (1992) Penetration of sized particle barriers by field populations of subterranean termites (Isoptera: Rhinotermitidae). *Journal of Economic Entomology* 85, 2275–2278.

Su, N.Y. and Scheffrahn, R.H. (1996) Evaluation criteria for bait-toxicity efficacy against field colonies of subterranean termites – a review. In: Wildey K.B. (ed.) *Proceedings of the International Conference on Insect Pests in the Urban Environment.* Edinburgh, UK, pp. 443–447.

Su, N.Y., Scheffrahn, R. and Weissling, T. (1977) Termite species migrates to Florida. *Pest Control* 65, 27–29.

Su, N.Y., Ban, P.M. and Scheffrahn, R.H. (1991) Evaluation of twelve dye markers for population studies of the Eastern and Formosan subterranean termite (Isoptera: Rhinotermitidae). *Sociobiology* 19, 349–362

Su, N.Y., Scheffrahn, R.H. and Ban, P.M. (1993) Barrier efficacy of pyrethroid and organophosphate formulations against subterranean termites (Isoptera: Rhinotermitidae). *Journal of Economic Entomology* 86, 772–776.

Su, N.Y., Tokoro, M. and Scheffrahn, R.H. (1994) Estimating oral toxicity of slow, acting toxicants against subterranean termites (Isoptera: Rhinotermitidae). *Journal of Economic Entomology* 87, 398–401.

Suzuki, K. (1991) Laboratory trial of biological control agents against subterranean termites. *International Research Group on Wood Preservation, Document No. IRG/WP/1475.* Twenty-second Annual Meeting, Kyoto, Japan.

Taguchi, S. and Yakushagi, T. (1988) Influence of termite treatment in the home on the chlordane concentration in human milk. *Environmental Contamination and Toxicology* 17, 65–71.

Tamashiro, M., Yates, J.R., Ebesu, R., Yamamoto, R.T., Su, N.Y. and Bean, J.N. (1989) Dursban TC insecticide as a preventive treatment for Formosan subterranean termite in Hawaii. *Down to Earth* 42, 1–5.

Tamashiro, M., Yates, J.R., Yamamoto, R.T. and Ebesu, R. (1991) Tunneling behaviour of the Formosan subterranean termite and basalt barriers. *Sociobiology* 19, 163–170

Thomas, C.R., Barlow, R.A. and Robinson, W.H. (1993) Dispersal of a termiticide foam beneath concrete slabs. *Japanese Journal of Sanitary Zoology* 44, 335–339.

Thoms, E.M. and Sprenkel, R.J. (1996) Overcoming customer resistance to innovation: a case study in technology transfer from the developer to pest control operator. In: Wildey, F.B. (ed.) *Proceedings of the International Conference on Insect Pests in the Urban Environment.* Edinburgh, UK, pp. 459–465.

Thorne, B.L. and Haverty, M.I. (1991) A review of intracolony, intraspecific and interspecific agonism in termites. *Sociobiology* 19, 115–145.

Thorne, B.L., Cohen, E.R., Forschler, B.T., Briesh, N.L. and Traniello, J.F.A. (1996) Evaluation of mark release capture methods for estimating forager population size of subterranean termite (Isoptera: Rhinotermitidae) colonies. *Environmental Entomology* 25, 939–951.

Tiben, A., Pearce, M.J., Wood, T.G., Kambal, M.A. and Cowie, R.H. (1990) Damage to crops by *Microtermes najdensis* (Isoptera: Macrotermitinae) in irrigated semi-desert areas of the Red Sea Coast. 2. Cotton in the Tokar Delta region of Sudan. *Tropical Pest Management* 36, 296–304.

Tshuma, J., Logan, J.W.M. and Pearce, M.J. (1991) Termites attacking field crops, pasture and forest trees in Zimbabwe. *Zimbabwe Journal of Agricultural Research* 26, 87–97.

Tsunoda, K. (1991) Termite bioassays for evaluation of wood preservatives. *Sociobiology* 19, 245–255.

Tsunoda, K. (1993) Termiticidal efficacy of synthetic pyrethroids 2. Effect of accelerated ageing on their termiticidal performance. *Wood Protection* 2, 67–73.

Tsunoda, K. and Nishimoto, K. (1986) *Japanese standardised methods for testing effectiveness of chemicals against termite attack. Document No IRG/WP/1290.* The International research group on wood preservation. Paper prepared for the 17th annual meeting, France, 25 May, 30, 20 pp.

Varma, A., Bala, K.K., Jaishree, P., Saxena, S. and König, H. (1994) Lignocellulose degradation by microorganisms from termite hills and termite guts, a survey on the present state of the art. *Federation of European Microbiological Societies Microbiology Reviews* 15, 9–28.

Veivers, P.C. Muhlemann, R. Slaytor, M., Leuthold, R.H. and Bignell, D.E. (1991) Digestion, diet and polyethism in two fungus growing termites *Macrotermes subhyalinus* Rambur and *Macrotermes michaelseni* Sjöstedt. *Journal of Insect Physiology* 37, 675–682.

Waller, D.A. (1996) Ampicillin, tetracycline and urea as protozoicides for symbionts of *Reticulitermes flavipes* and *R. virginicus* (Isoptera: Rhinotermitidae). *Bulletin of Entomological Research* 86, 77–81.

Watanabe, H. and Noda, H. (1991) Small scale rearing of a subterranean termite *Reticulitermes speratus* (Isoptera: Rhinotermitidae). *Applied Entomology and Zoology* 26, 418–420.

WHO (1989) International programme on chemical safety (IPCS). *Aldrin and Dieldrin.* Geneva, Switzerland. World Health Organization, Environmental Criteria, No. 91.

Williams, R.M.C. (1959) Colony development in *Cubitermes ugandensis* Fuller (Isoptera: Termitidae). *Insectes Sociaux* 6, 291–304.

Williams, R.M.C. (1973) Evaluation of field and laboratory methods for testing termite resistance of timber and building material in Ghana with relevant biological studies. *Tropical Pest Bulletin* 3. Centre for Overseas Pest Research. Overseas Development Administration, Foreign and Commonwealth Publication UK, 64 pp.

Williams, R.M.C. (1976) Factors limiting the distribution of the building drywood termites (Isoptera, *Cryptotermes* spp.). *Material und Organismen* 3, 394–406.

Williams, R.M.C. (1977) Ecology and physiology of structural wood destroying Isoptera. *Material and Organismen* 12, 111–140.

Wiseman, S. and Eggleton, P. (1994) *The Termiticide Market* (D588). Agrow PJB Publications Ltd, Richmond, Surrey.

Wood, T.G. (1977) Food and feeding habits of termites. In: Brian, M.V. (ed.)

Production Ecology of Ants and Termites. International Biological Programme 13. Cambridge University Press, Cambridge, pp. 55–80.

Wood, T.G. (1981) Reproductive isolating mechanisms among species of *Microtermes* (Isopotera: Termitidae) in the Southern Guinea Savanna near Mokwa, Nigeria. In: Howse, P.E. and Clement, J.L. (eds) *Biosystematics of Social Insects.* Academic Press, London, pp. 309–325.

Wood, T.G. and Cowie, R.H. (1988) Assessment of on-farm losses in cereals in Africa due to soil insects. *Insect Science and its Application* 9, 709–716.

Wood, T.G. and Pearce, M.J. (1991) Termites in Africa: the environmental impact of control measures and damage to crops, trees, rangeland and rural buildings. *Sociobiology* 19, 221–234.

Wood, T.G. and Sands, W.A. (1977) The role of termites in ecosystems. In: Brian, M.V. (ed.) *Production Ecology of Ants and Termites.* International Biological Programme 13. Cambridge University Press, Cambridge, pp. 245–392.

Wood, T.G. and Thomas, R.J. (1989) The mutualistic association between Macrotermitinae and Termitomyces. In: Walding, N., Collins, N., Hammond, P.M. and Webber, J.F. (eds) *Insect–Fungus Interactions.* Academic Press, London

Wood, T.G., Bednarzik, M. and Aden, H. (1987) Damage to crops by *Microtermes najdensis* (Isoptera, Macrotermitane) in irrigated semi-desert areas of the Red Sea Coast. 1. The Tihama region of the Yemen Arab Republic. *Tropical Pest Management* 33, 142–150.

Woodrow, R.J. and Grace, J.K. (1977) Cooking termites in the Aloha State. *Pest Control* 65, 57–62.

Yaga, S. (1972) On the secretion of the termite *Coptotermes formosanus* Shiraki; the components of sugars and amino acids in the secretion of workers. *Science Bulletin, College of Agriculture.* University of Ryukus, Okinawa, Japan 19, 481–488.

Yamin, M.A. (1979a) Flagellata with lower termites. *Sociobiology* 14, 91.

Yamin, M. (1979b) Scanning E.M. of some symbiotic flagellates from the termite *Zootermopsis. Transactions of the American Microscopical Society* 98, 2.

Yoshimura, Y. and Tsunoda, K. (1994) Alternative protection of Japanese houses from subterranean termite invasion In: Lenoir, A., Arnold, G. and Lepage, M. (eds) *Termites in Urban Areas.* Les Insectes Sociaux, Paris, Sorbonne, August (1994) Abstracts 12th International Congress International Union for the Study of Social Insects, p. 259.

Yoshimura, T., Tsunoda, K. and Takahashi, M. (1994) Cellulose metabolism of the symbiotic protozoa in the termite *Coptotermes formosanus* Shiraki (Isoptera: Rhinotermitidae) IV. Seasonal changes of the protozoan fauna and its relation to wood-attacking activity. *Mokuzai Gakkaishi* 40, 853–859.

Zoberi, M.H. and Grace, J.K. (1990) Isolation of the pathogen *Beauveria bassiana* from *Reticulitermes flavipes* (Isoptera: Rhinotermitidae). *Sociobiology* 16, 289–296.

Index

abdomen 14, 58
aggression 55–60, 151
agriculture 65, 96–99
alate 10, 14, 65, 86
aldrin 101, 105–106
ants 1, 87
arsenic compounds 106, 108, 112
ash 104, 106

bacteria 51–52, 149
baits 109–113, 139–141
barriers 38, 114–116
beetles 74–75, 88
behaviour
  communication 40–46
  defence 55–60
  feeding 46–53
  foraging 60–63
  nest building 63–64
  water requirements 53–55
benefits
  methane (non-beneficial role) and other gases 80–81
  minerals 79–80
  nitrogen 79
  water infiltration 80
biological control 118–120
birds 76, 86
boric acid 106, 108, 112
building
  galleries 60–61
  nest building 63–64
buildings
  as pests 94–96
  baiting 109–113
  control soil application 104–107
  socioeconomic considerations 107
  wood treatment 108–109

cables 96, 148
calcium 79–80
cannibalism 53, 136
carbofuran 103, 106
castes
  characteristics 1–2
  replacement reproductives 16–19
  reproductives 10, 14–15
  royal pair 15–16
  soldiers 10–13
  workers 9–10
classification
  behavioural 30, 31
  external morphology 19–20, 21–28
  feeding/nest/chemical 21
  internal features 29, 30
collections
  collecting methods 123–124
  identification 124
  sending off specimens 125–126
  storage of collections 125
chemical control
  buildings 104–113
  crop protection 101–103
  forestry 103
  plants and plant extracts 113
  rangeland 103–104
  residues and contamination 113–114
  stored product protection 104
chlordane 101, 106, 113
chlorpyrifos 102–103, 106
colony foundation 15, 65

communication
    broodcare 46
    grooming 45–46
    pheromones 44–45
    sense organs 41–44
control
    biological 118–120
    chemical 101–114
    future predictions 121–122
    physical and cultural 114–118
    safety 120–121
cotton 98, 138–139
creosote 108
crops 96–99, 101–103, 115–116
culture
    containers 128
    culture rooms 127–128
    problems 135–136
    setting up colonies 128–135

damage
    assessment 100
    buildings 94–96
    crops 96–99
    detection 99–100
    non-cellulose materials 96
    timber 93–94
    trees 99
date palms 74, 99
defence
    chemicals 57–60
    head shape 56–57
    mandibles 56
    natural defences 58–59
    worker defence 58
detection 99–100
development 1, 18
dieldrin 101, 106, 113
diet 46–50, 148
distribution
    factors affecting 37–39
    in foraging 62
    islands 37–38
    land bridges 37
    natural barriers 38
    pest 33–37
    spread by humans 39
    world 32–33
dung 48
dusts 106, 108
dyes 145, 152

ecology
    benefits to the environment 79–81
    environmental factors 81–85
    predators and parasites 85–90
    refuse 90
    soil type 77
    uses of mounds 90
    vegetation types 78–79
environmental factors
    food availability 84–85
    humidity 82
    predators and parasites 85–90
    rainfall 82
    salt tolerance 82
    temperature 83–84
enzymes 50
evolution
    fossil evidence 2
    geological time scale 8
    relationship to cockroaches 7–9

feeding
    classification 30
    contact poisons 140–141
    enzymes and digestion 50
    feeding experiments 144–149
    food availability 84–85
    food preferences 91–93
    fungus-comb consumption 48–50
    nitrogen sources 52–53
    other food sources 48
    protozoa 50–52
    trophallaxis 46–47
    water requirements 53–54
    wood feeding 47–48
field testing 103–104, 109–113, 137–141
fipronil 103, 106
fontanelle 10, 11, 23, 57–58
food preferences
    crops 93
    wood 92
foraging
    distribution 62
    galleries 60–62
    monitoring 139–140
    trail chemicals 63
    trails 62–63
fossils 2, 8–9
fumigation 109, 120
fungicide 112
fungus-combs 15, 48–50, 71–74, 143

galleries 60–64, 67–70
grooming 45–46, 143–144
groundnuts 97
growth regulators 110
gut 20, 29, 51

harvester termites 24, 36, 62, 78, 84, 98
heat treatment 109

identification 12, 19–31, 124–125
imago *see* queen
imidaclorprid 102–103, 106
insecticides 106–107

life-cycle 1, 18
litter 48, 70–71

magnetic effects 46
maize 62, 93, 97
mark–recapture 142
markers 145, 152
measurement 21
methane 80
methyl bromide 109, 120
mites 86, 136
monitoring methods
  foraging 139–141
  plant mortality 137–139
  sampling 137
  value of mark recapture methods 142
  viability of field trials 141
morphology 2, 9–31, 41–44, 55–57, 124–125
mounds 67–70, 116–117
mouthparts 11–12, 21–27, 41–42, 47–48, 56

nematodes 118, 149–150
nests
  chemical control 104
  classification 30
  food storage 70–71
  foundation 65
  fungus-combs 71–74
  mounds 67–68
    removal 116
  nest/mound collection 133
  predators and parasites in 86–88
  subterranean termites 67
  ventilation 68–72
  wood-dwelling termites 66
nitrogen 52–53, 114–117
non-chemical control (physical and cultural)
  barriers 114–115
  crops 115–116
  mound removal, flooding and local methods 116–117
  plant protection tips 116
  resistant timbers 117
nymphs 2, 9, 18

organochlorines 101, 105–106, 113–114

parasites 85–90, 136
pastures 78, 84, 98
pathogens 118–120
pests
  assessment 100
  damage recognition and detection 93–100
  distribution 33–37
  food preferences 91–93
pheromone 18–19, 63
populations 13, 85 139–142
predators
  of alates 85–86
  food for man and domestic animals 88–89
  foraging termites 88
  termites inside nests 86–88
preservatives 108
pressure impregnation 108
protozoa 31, 50–52, 151
pyrethroids 105–106, 108

queen 10, 14–18, 45–47, 54, 67–68, 88–89, 116, 133–134

repellents 44–45
reproductives *see* queen
residues 113–114
resistance 92–93, 117–118
rubber 96, 148
runways *see* galleries

safety 120–121
sampling methods 137–142

sense organs 41–44
sexing 14
soil treatment 105–107
soils 38, 68, 77, 82, 151
soldiers 10–14, 19–20, 56–58, 64, 87, 133
sugar cane 21–25, 42, 93, 97, 102
swarming 14, 30, 82, 86, 89–90, 134

taxonomy 10–13, 19–31, 123, 126
teak 92
Termitomyces 49
termitophiles 74–76
    invertebrates 74–75
    vertebrates 76
testing
    biocontrol agents 149–150
    contact poisons 143–144
    feeding
        on treated papers (tolerance and preference) 144–145
        on wood, treated or resistant 145–147
    other techniques and procedures 150–152
    testing other materials 147–149
timbers 99, 108–109
trails 15, 62–63, 150–151
traps 109–113
trials 137–142
trophallaxis 46–47, 145
tunnelling 60–62, 93–100

vegetation
    crop damage 96–99
    deserts 79
    primary and secondary forests 78–79
ventillation 68–70
virus 120
vision 16, 43

water 53–55, 80, 82, 95–96
wings 14–15, 20, 26–28
wood 66, 78–79, 92–94, 108–109

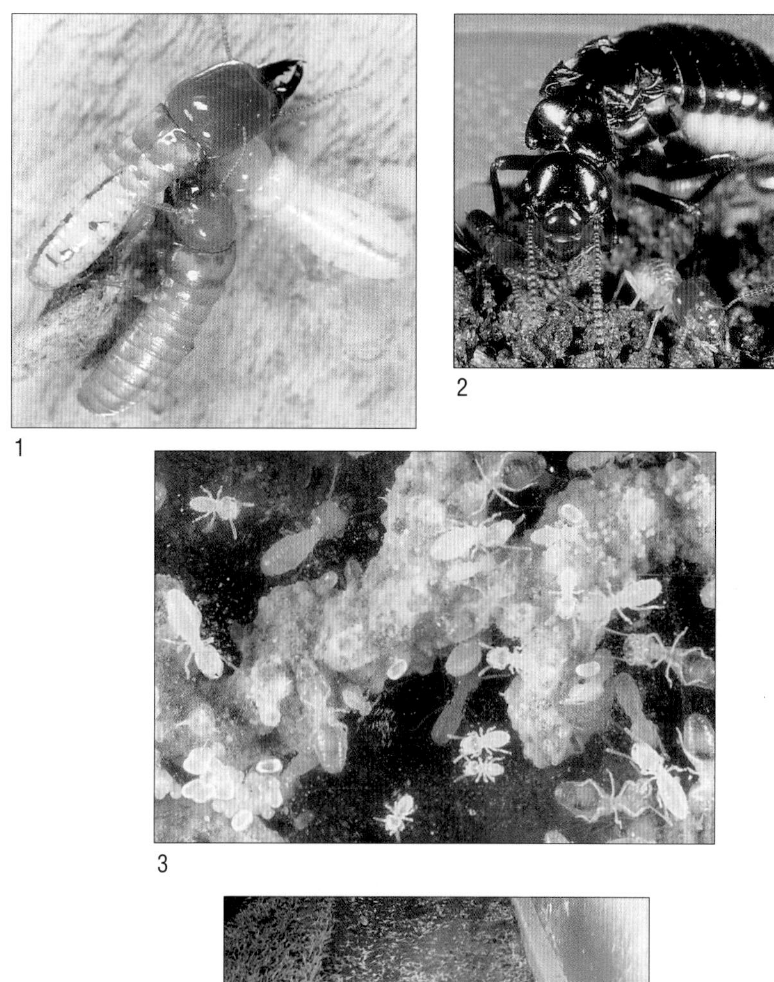

1

2

3

4

Plate 1. Soldier, worker and supplementary reproductive of *Neotermes.*
Plate 2. Young queen of *Pseudocanthotermes* with a new worker.
Plate 3. Workers, larvae and eggs of *Microtermes* in a chamber containing a fungus-comb covered in asexual spores.
Plate 4. Thousands of cast-off wings left after a flight of *Macrotermes* alates the previous evening.

5 7

6

Plate 5. Soil sheeting built over the trunk of a fruit tree by *Odontotermes.*
Plate 6. A nest of *Neotermes* showing chambers and tunnels made inside a piece of wood, which is also their food source.
Plate 7. Mushroom-shaped rain shields built by *Cubitermes* against the side of a tree in the People's Republic of the Congo.

Plate 8. Carton nest built by *Nasutitermes* on the side of a tree in Australia.
Plate 9. Nest of *Coptotermes* built inside a telephone cable box in Hong Kong.
Plate 10. A network of subterranean tunnels in soil made by *Microtermes* and leading to chambers containing fungus-combs.
Plate 11. A large spiral plate with salt-coated vanes built under the base of the nest by a species of *Macrotermes* in Africa.
Plate 12. A mound of *Trinervitermes* from South Africa opened up to show the many chambers, some of which contain stored grass (right-hand side).

13

15

14

Plate 13. The central hive of an African *Macrotermes* mound showing the area occupied by fungus-comb.
Plate 14. Development of shrubs and bushes on the nutrient-rich soil of an abandoned *Macrotermes* mound.
Plate 15. A Ugandan farmer setting up an alate trap for *Odontotermes*, made of sticks which will eventually be covered with dead grass.

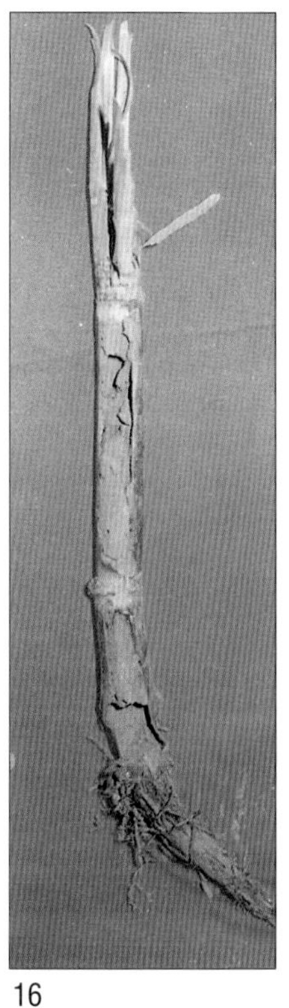

16

17

18

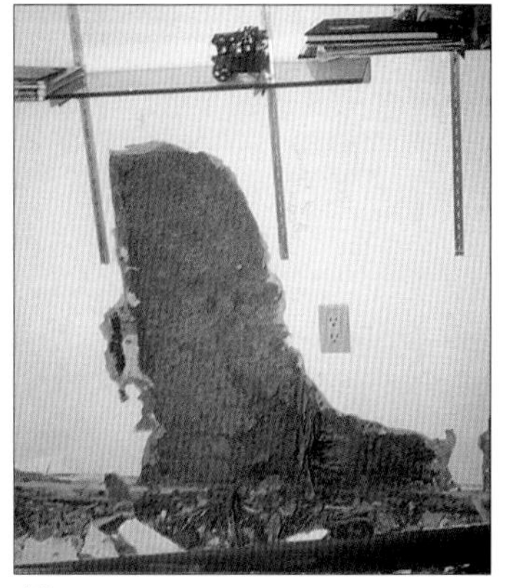

19

Plate 16. A stem of sugar cane hollowed out by termites in Pakistan.
Plate 17. Subterranean termite attack on a telephone pole in northern Sudan.
Plate 18. Southern USA Yellow Cypress attacked by termites. The lower unattacked pieces are treated heartwood pine.
Plate 19. Attack on an internal wall structure by *Coptotermes* in the USA.

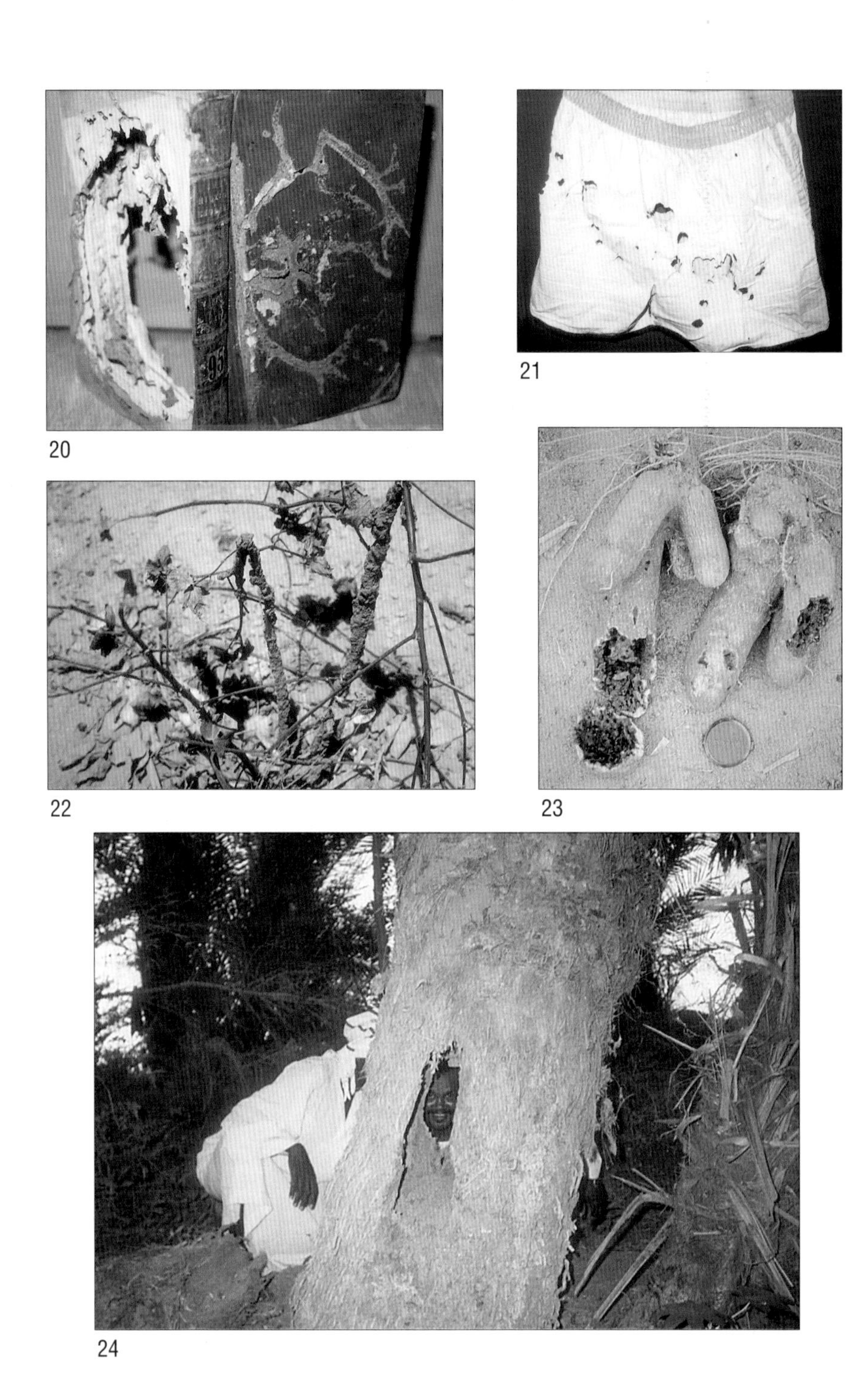

Plate 20. Valuable books are eaten by termites.
Plate 21. Sports shorts damaged by termites entering a store house.
Plate 22. Sudanese cotton plants destroyed by *Microtermes.*
Plate 23. Yams totally eaten out by *Amitermes* in Nigeria.
Plate 24. The trunk of a date palm from northern Sudan eaten out by *Odontotermes.*

25

28

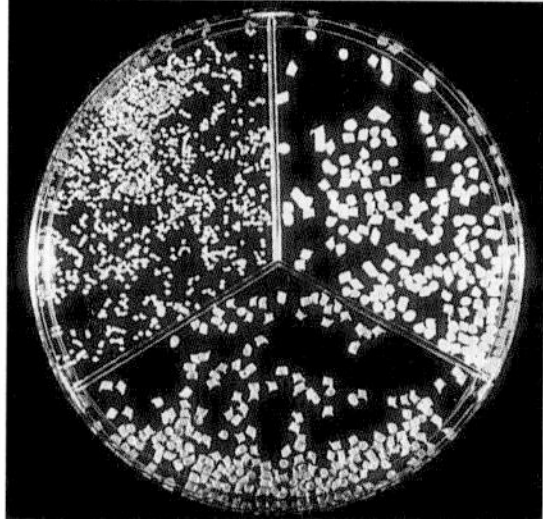

26

27

29

Plate 25. A fruit tree damaged by *Odontotermes* in Pakistan.
Plate 26. Slow release granules for termite control.
Plate 27. Injecting termiticide into holes drilled through concrete as an after treatment against termites.
Plate 28. A fumigation tent covering a home in the USA.
Plate 29. A method of baiting using a PVC pipe containing a food source which has been placed in a *Coptotermes* mound in Australia.

30

31

32

Plate 30. A field method using cardboard disc baits for monitoring foraging and the effect of control agents.
Plate 31. A box bait method used in Australia to attract and treat termites in order to destroy nests both outside and inside the home.
Plate 32. A dead termite releasing new infective parasitic nematodes. These have to enter other termites in order to become effective for termite control.